Critical and scientific thinking

Critical & Scientific Thinking

A Short Introdcution

Michael Vlerick

Open Press Tilburg University

ISBN 978-9-40-366878-9
DOI 10.26116/IHOX-9K96

Design and Typesetting by Thomas F. K. Jorna
Cover by Mor Lumbroso
Copyediting by Meike Robaard
Coordination by Violet S. Zagt

Publishers of Trial and Error

Published by Open Press Tilburg University,
Tilburg, The Netherlands
https://www.openpresstiu.org
info@openpresstiu.org

Contents

	Preface		I
	1	What is this book about?	I
	2	Who is this book for?	I
	3	Why read this book?	II
1	**What is Philosophy of Science?**		I
	On the importance of philosophy of science		
	1	What is science?	I
	2	What is philosophy?	4
	3	The importance of philosophy of science	5
	4	The structure of this book	6
	5	Summary	7
	6	Further reading	8
2	**Predictably Irrational**		9
	An overview of common reasoning errors		
	1	What is (and what is not) critical thinking?	9
	2	The goal of critical thinking	10
	3	The usefulness of critical thinking	10
	4	The tenacity of nonsense	12
	5	Three rules of thumb	13
	6	Predictably irrational	15
	7	Problem 1: The 'Linda' problem	15
	8	Problem 2: The base rate fallacy	16
	9	Problem 3: The base rate fallacy	16
	10	Problem 4: The birthday problem	17
	11	Problem 5: Exponential reasoning	17
	12	Problem 6: Exponential reasoning	17
	13	Problem 7: The availability bias	19

14 Problem 8: The availability bias 19
15 Problem 9: The availability bias 19
16 Problem 10: Anchoring 20
17 Problem 11: The framing effect 21
18 Problem 12: The Allais paradox 21
19 Problem 13: The hindsight bias 22
20 Problem 14: The confirmation bias 23
21 Problem 15: Self-overestimation 24
22 Summary 25

3 **Why Are We Irrational?** 27
The evolutionary origin of irrationality
1 The peculiar architect of our thinking 27
2 Evolution by natural selection 27
3 What does this entail for our thinking? 28
4 Truth is an expensive means to an end 29
5 System 1 and system 2 30
6 The fallibility of system 1 31
7 Taking stock 34
8 Other sources of irrationality 35
9 Summary 40
10 Further reading 41

4 **Irrationality in Action** 43
How reasoning errors lead to domains of irrationality
1 Superstition, horoscopes, and palm reading 43
2 Correlation does not imply causation 44
3 Order in randomness and randomness in order 45
4 Chance blindness 45
5 Causal reasoning errors 46
6 Conspiracy theories 47
7 Pseudoscience 49
8 Religion 55

	9	The myth of 'Homo economicus'	57
	10	Summary	59
	11	Further reading	60
5	**Mastering Critical Thinking**		61
	How we can guard our thinking against reasoning errors		
	1	Three sources of reasoning errors	61
	2	Intuitive reasoning errors	61
	3	Emotions	68
	4	How can we curb the confirmation bias?	70
	5	Summary	75
6	**The Importance of Critical Thinking**		77
	1	Are there beneficial illusions?	77
	2	The impact of irrationality on the world	78
	3	What about religion?	80
	4	Critical thinking and moral progress	82
	5	Critical thinking and progress in general	85
	6	The major challenges of today	86
	7	A lasting struggle	87
	8	Summary	88
7	**The Importance and Reliability of Science**		89
	On the scientific method and the demarcation criterion		
	1	The scientific method	89
	2	Human and natural sciences	90
	3	A self-correcting process	92
	4	How is science protected against the reasoning errors of scientists?	93
	5	The importance of scientific progress	95
	6	Self-censorship in the human sciences	96
	7	The demarcation criterion	97

8 Sokal's hoax 101
9 Striking a balance 102
10 Summary 104

Appendix 105
11 List of reasoning errors 105
12 Detect the reasoning errors 107
13 Answers 113

Bibliography 118

Preface

WHAT IS THIS BOOK ABOUT?

This book is about critical and scientific thinking. You will learn what reasoning errors we all tend to make, why we make those reasoning errors, what they lead to, and how we can improve our thinking. Finally, you will learn about the importance of critical and scientific thinking, and what characterizes the sciences, i.e. what makes them distinctive (from pseudoscience and non-scientific domains of inquiry) and reliable.

WHO IS THIS BOOK FOR?

This book is written for students taking a course on philosophy of science or critical thinking. The book provides content for seven lectures (the seven chapters) and one seminar (appendix). It can stand on its own for courses consisting of seven lectures with a study load of 3 ECTS or might be combined with a historical or thematic overview of the philosophy of science (e.g. Dooremalen et al., 2021) for courses consisting of fourteen lectures with a study load of 6 ECTS.

I use this book as part of my teaching of philosophy of science for students at Tilburg University. Hopefully, it will also find its way to other lecturers.

Ultimately, I hope that this book finds its way out of academia to reach a wider audience. Critical thinking concerns everyone. It is of great value on both a personal and societal level. Indeed, as I discuss in Chapter 6, critical thinking is a driver of progress, both with respect to knowledge and innovation as with regards to morality.

WHY READ THIS BOOK?

Critical thinking is one of the biggest hiatuses in our education system. Learning to distinguish sense from nonsense is of great importance in the information age that we live in. In a systematic way, this book helps you to gain insight into, and subsequently eliminate, the most important reasoning errors that we all tend to make. It also helps you to debunk weak and fallacious arguments and unreliable information.

In addition to understanding what critical and scientific thinking entails, you will learn more about what makes science reliable. In times of skepticism regarding science, where (sometimes dangerous) pseudoscientific and conspiracy theories run rampant, this is particularly important.

Critical thinking, as I conclude in Chapter 6, is not a matter of intellectual preference or even self-interest (although one certainly benefits from thinking critically). It is first and foremost a matter of moral and social responsibility. Better thinking leads to a better world. With this book I hope to contribute to that important goal and you, dear student or reader, can do the same!

Enjoy your reading!

1 What is Philosophy of Science?

On the importance of philosophy of science

WHAT IS SCIENCE?

The central question that philosophers of science ask is: What is science? This question may seem easy to answer at a first glance. The sciences are physics, chemistry, biology, etc., and not music, art and religion. This, of course, is true, but it raises the very same question: What distinguishes those scientific domains and activities from non-scientific domains and activities? What characteristics do the sciences share with each other and not with non-scientific domains? What, in other words, makes science science? (Okasha, 2002).

This question, too, may seem easy to answer. Sciences attempt to explain certain aspects of reality based on observations. But whilst that is certainly not a bad answer, it is not entirely satisfactory. Astrology (horoscopes), too, seeks to explain aspects of reality based on observations, as does religion. So, what is it that demarcates the sciences as science? What is it that distinguishes science from so-called pseudoscience? The latter refers to theories and practices that may appear scientific but are not (such as astrology, creationism, and certain forms of alternative healthcare). Finally, we must also ask what characteristics make science reliable or – in any case – more reliable than pseudoscience. These are the questions we will address and answer in this book.

From the question of what science is, however, follows a series of other questions that philosophers of science ask. What is the relationship between scientific theories and reality? Realists, for example, think that scientific theories represent reality truthfully or, at least that they can represent reality truthfully. Anti-realists disagree. According to the latter, we can only claim that scientific theories can make accurate predictions, not that they actually represent reality (i.e. that they are faithful depictions of the reality they describe).

Another question often asked by philosophers of science is: How do the sciences evolve? Contemporary scientific theories are often quite different from those of, say, the nineteenth century. How did this change come about? According to the philosopher of science Karl Popper (1963) - who will be discussed later – scientific change happens in a gradual way. New theories are typically revisions of previous theories, and we may therefore assume that, in general, the sciences are improving over time. They represent the world more truthfully than the theories they replaced.

Another prominent philosopher of science, Thomas Kuhn (1970), objects. Sciences undergo 'revolutions', according to Kuhn, discarding just about everything that came before it. This debate, of course, also has important implications for the question of scientific realism. Someone like Kuhn joins the ranks of the anti-realists because he sees science as an intellectual activity within a so-called paradigm. A paradigm is built on basic assumptions for which there is no evidence. In case of a revolution, one simply discards the old paradigm and starts again in a new paradigm. Therefore, according to Kuhn, the sciences do not come closer to the truth over time, they just switch ways of looking at world (and one way is not inherently more truthful than another way).

You may wonder what use all this is to practicing specific sciences. That is a valid point. The discussion between realists and anti-realists changes little to nothing for the way scientific investigation is conducted. But philosophers of science are not just concerned with science in general, they also think about specific sciences. There is a philosophy of physics, a philosophy of biology, a philosophy of psychology, and a philosophy of economics. In the philosophy of economics, for example, philosophers question whether economic models objectively describe economic reality. Perhaps subjective values creep in? For instance, the value that economists place on freedom (and free entrepreneurship) may lead them to be slightly biased towards perceiving (free) market mechanisms as efficient (and perhaps turn a blind eye to shortcomings of the free market). Or the opposite effect may occur for economists who take moral issue with the inequalities produced by an unregulated market.

This subjectivity, some say, is inevitable because economists – like scientists in any other field – are necessarily selective in what they will measure and represent. We can never represent the economy – with all of its complexity and idiosyncrasies – in its actual entirety, and so we focus on certain aspects and relationships that we (often unconsciously) find important. The same is true in other sciences. Scientists are constantly making choices about what to study, what causal relationships to uncover, etc. The choice of investigating certain features and connections rather than others, comes from value judgments (what scientists believe is important and relevant). Therefore, according to some – contrary to what is often assumed – science can never be completely objective.

Furthermore, philosophers of science who are concerned with a particular science often engage in critical reflection on the assumptions scientists make within their domain. For example, models in classical economics assume that economic agents (consumers and investors) will rationally maximize their "utility' (pleasure, happiness, value). For example, the consumer is assumed to make a rational cost-benefit analysis when faced with the choice of whether or not to buy a good. She will, according to this view, only purchase something when she cannot derive more utility from another purchase for the same price. In recent decades, however, that view of the rational economic actor has been undermined by research in so-called behavioral economics. It turns out that we are not the rational actors that we are assumed to be by the economic models. This has important implications (we will return to this later).

Unlike theories in, say, astrophysics, economic theories - and other theories in the human sciences - have an important impact on our lives. The reason for this is simple: economic theories and models inform economic policy. Misconceptions in the economic sciences lead to mismanagement of society. Some say the responsibility for the global economic crisis of 2008 lies in part with the flawed models that economists created. As a result, banks and other financial institutions considered certain complex financial instruments safe because the models indicated it as such. In retrospect, these models turned out to be widely off mark in their risk assessment. The same is true in psychology and the social sciences. Bad theories lead to bad practices (e.g. in psychotherapy) and bad policies. So, we must permanently question the theories and models we use, as well as the assumptions on which these models are based. After all, our well-being depends on it.

WHAT IS PHILOSOPHY?

This brings me to the question of what exactly philosophy is (and what its purpose is). With regards to that question, there are as many answers as there are philosophers. Answering the question 'what is philosophy?', is actually itself a branch of philosophy: the so-called 'philosophy of philosophy'. I will not go into lengthy digressions about philosophy, nor will I give you a precise definition. But it is useful to straighten out a series of misconceptions.

Practicing philosophy does not mean developing obscure theories in an ivory tower. Nor does it involve poetic reflections on the meaning of life. Or at least, it should not. Rather, philosophy is a way of thinking. First, it is a rational way of thinking. In the West, philosophy emerged roughly 2,500 years ago when, for the first time in history, people tried to understand the world without resorting to mythological and religious stories, but by using their own power of understanding. Sciences are also rational, and this is no coincidence: the modern sciences emerged from philosophy. The first modern scientist, Isaac Newton, considered himself a natural philosopher. The father of economics, Adam Smith, was also a philosopher.

Second, philosophy is critical. It takes nothing for granted, but questions everything. Here, it differs in an important way from the sciences. Scientists are also critical and will subject theories to empirical tests before accepting them, but they generally do not question the basic assumptions of their science. Philosophers, however, do. Unlike the sciences, therefore, philosophy is radically critical. It questions the grounds or the foundations (radix is the Latin term for 'root') of any theory.

Finally, the reach of philosophy is much broader than that of the individual sciences. The sciences have well-defined domains. Economics, for example, is concerned with the distribution of scarce resources (products and services) in society. Psychology is concerned with human thought and behavior and their underlying mental processes. This is not the case for philosophy. Philosophy does not confine itself to a particular domain. It looks beyond the boundaries of different domains. Doing so, philosophy can develop a different perspective on certain issues. It can freely combine insights from different scientific domains to arrive at new insights. For example, it can combine insights from biology, psychology, economics, sociology, and anthropology to think about how we can best organize society and how we can best address certain social problems. (For the Dutch speaking among you, if you are interested, I attempted to do so in my book 'De tweede vervreemding', on globalization and the prospect of global cooperation – Vlerick, 2019).

THE IMPORTANCE OF PHILOSOPHY OF SCIENCE

The importance of philosophy in general and of philosophy for the sciences in particular, lies precisely in its reflective and critical approach as well as in its broad scope. In this way it both assists and supplements the sciences. By reflecting on scientific theories, it helps to clarify important scientific concepts. It sheds a critical light on unfounded assumptions in scientific fields. And important findings from the empirical sciences, in turn, are combined and situated in a broader context.

Finally, the philosophy of science is particularly attentive to the processes of the sciences. Too often the emphasis is put only on what the sciences tell us about the world, not on how the sciences arrived at those insights. Philosophy of science accentuates not only what appears on your plate (the ready-made theories), but also what happens in the kitchen (how those theories come about). This is necessary to understand what characterizes science and what makes sciences reliable.

THE STRUCTURE OF THIS BOOK

In a typical philosophy of science course, the question of what science is, is addressed by providing a historical or thematic overview of what prominent philosophers of science have said about it. In this book, however, I take a different approach. I am taking on the question of what science is and what makes it reliable from the perspective of critical thinking instead. By first gaining insight into our own thinking - how our thinking systematically misleads us and how we can improve our thinking - it becomes clear how the scientific context and methodology protect against reasoning errors and generally lead to increasingly reliable theories.

In the next chapter ('Predictably Irrational'), you will find out exactly what 'critical thinking' is, and I will show you, through a series of entertaining riddles, that your spontaneous thinking is misleading in predictable ways. In Chapter 3 ('Why are we Irrational?'), I will explain why this is the case. In Chapter 4 ('Irrationality in Action'), I will explain how those reasoning errors lead to certain forms of irrationality, such as superstition, conspiracy theories, pseudoscience, and religion. In Chapter 5 ('Mastering Critical Thinking'), I will introduce you to the remedy. You will learn how to avoid reasoning errors and how to think (more) critically. In Chapter 6 ('The Importance of Critical Thinking'), I will discuss why this important. Finally, in Chapter 7 ('The Importance and Reliability of Science'), I will link these insights back to the questions with which I started this book, namely: What demarcates science and what makes sciences reliable? In the Appendix, you will find material to practice your critical thinking skills. You will find a list of the most important reasoning errors, a series of case studies in which reasoning errors feature for you to detect, and the answers to these case studies. It is best to go over the contents of the Appendix after you have read Chapter 2 and before you begin Chapter 3.

SUMMARY

What is the central question of the philosophy of science?

'What is science?'

What questions follow from this central question?

- What distinguishes science from pseudoscience and from non-scientific fields?
- What is the relationship between scientific theories and reality?
- How do the sciences evolve?

What is philosophy?

A way of thinking:

- Rationally
- Radically critical
- With broad scope

What is the importance of philosophy for science?

- To clarify concepts
- To critically evaluate assumptions
- To put findings in a broader context

FURTHER READING

Okasha, S. (2002). *Philosophy of science: A very short introduction.* Oxford University Press

Predictably Irrational

An overview of common reasoning errors

WHAT IS (AND WHAT IS NOT) CRITICAL THINKING?

The term 'critical thinking' is often used, but it is not always clear what exactly it refers to. So, what is critical thinking? Critical thinking is rational thinking. It aims to generate justified beliefs (beliefs that we may assume to be true) by systematically analyzing the way in which beliefs have been formed. In other words, critical thinking means that we assess the reliability of our beliefs by reflecting on how these beliefs were formed. Moreover, critical thinking is also autonomous thinking. A critical thinker does not adopt beliefs simply because they are part of a cultural tradition or expressed by an authority figure. In short, critical thinking consists in forming beliefs in a rational (not intuitive and / or emotional) and autonomous way (not relying on tradition and/or authority).

What is critical thinking not? Critical thinking is not 'negative' thinking. It does not aim to undermine every claim. Critical thinking does not mean that we question everything permanently. It does not necessarily lead to skepticism: the position in which one suspends all beliefs by 'knowing that one does not know'. Critical thinking is not intelligent or creative thinking either. Sometimes intelligent thinking leads to very uncritical beliefs (think for example of ingenious conspiracy theories). Finally, critical thinking cannot simply be identified with well-informed thinking. Being well informed is a necessary condition for arriving at justified beliefs, for without good information we cannot come to a justified belief, but not a sufficient condition, because even with good information we can still come to unjustified beliefs. We can indeed misinterpret correct information.

THE GOAL OF CRITICAL THINKING

Critical thinking aims to distinguish sense from nonsense, good arguments from bad arguments, reliable thinking from unreliable thinking. To do so, we must focus on the source of our thinking: our thinking apparatus. By gaining insight into our own thinking, we are better able to estimate the reliability of the outcomes of such thinking. It is important to realize that we are not born with the ability to think critically. Critical thinking must be learned. In fact, critical thinking often goes against our spontaneous way of thinking. We must constantly beware of reasoning errors or fallacies.

No one - however intelligent they may be - is immune to irrational thinking. On the contrary, sometimes intelligent people are actually more susceptible to adopting irrational beliefs because they are better able to defend those views against counterarguments. Take Sir Arthur Conan Doyle, author of the Sherlock Holmes detective stories, for example. Doyle was led to believe that fairies existed by two teenage girls armed with their dad's camera and fairy dolls. Doyle would defend his outlandish view against sceptics and give complex arguments for the existence of spiritual entities.

THE USEFULNESS OF CRITICAL THINKING

Before going into how and why our minds lead us astray, and how we can guard our thinking against reasoning errors, we have one more important question to address: Why should we think critically? What is the use of critical thinking? Critical thinking is not a purely intellectual exercise. It has a real and important impact on our daily lives. We make many decisions every day. These decisions range from trivial ones about what to have for dinner, and if we should buy a new phone, to more important decisions such as what to study at university and which professional career to pursue. We make these decisions based on information. Information about the nutritional value, price, and taste of food products, about the price and quality of that new phone, or about the study curriculum and the profession we are considering.

This is relatively new in the history of humankind. Never before did or could we make as many decisions as today. Deciding which professional career to pursue, whom to marry, how many children to have, where to live and what to consume is a fairly recent phenomenon. In the Middle-Ages all these things were a given: almost everybody did what their father or mother did, was given away in marriage, did not do family planning, lived in their native village, and consumed what was available (given their social class). Life was already settled before it even began. Today, in modern societies at least, that is not the case.

Consequently, we have never been so dependent on information. And of information there is anything but a lack. Indeed, we are constantly flooded with it. The internet and other media bombard us daily with an endless stream of information. The problem, however, is that not all information is reliable and that we typically do not get an assessment of reliability with the fragments of information that reach us. We must find that out for ourselves. By now, most of us are aware that an email from an obscure billionaire who promises us a huge sum if only we help him unlock his heritage, is not exactly reliable. But so much misinformation still goes 'viral'. The internet is overflowing with unfounded health warnings against, for example, the use of microwave ovens or cellphones. We are also fed a constant stream of health and other advice that is far from reliable. From the next detox cure that will make us look ten years younger, to 'superfoods' of which we cannot eat enough or – and this has much more severe consequences – completely unfounded and alarmist claims about covid- and other vaccines. These are big claims, without (sufficient) evidence to back them up. Nonsense has always been around, but the amount of nonsense we are served today is greater than ever.

Moreover, nonsense breeds more nonsense. Beliefs do not emerge in isolation. Our worldview consists of a complex web of interwoven beliefs. This means that illusions or irrational and erroneous views tend to branch out in our thinking (Boudry, 2016). Anyone who believes in the predictive power of astrology will probably be more susceptible to other illusions such as believing in the existence of people with paranormal gifts, psychics, and the efficacy of treatments such as 'energy healing'.

THE TENACITY OF NONSENSE

Nonsense, as I have previously pointed out, is a historical constant. All eras and cultures have their irrational views. Interestingly, however, whereas blatantly irrational views generally seem completely absurd to an outsider, people within groups that hold these views are often not aware of the bizarre nature of their convictions. We do not have to go back far into the past to find seemingly absurd beliefs. In the 18th century, a large part of the population believed in witchcraft, in the fact that an English woman, Mary Toft, gave birth to rabbits and that there were recipes to produce not only gold (alchemy) but also live animals (such as a scorpion by placing basil leaves between two stones and letting it heat in the sun).

From the outside, these views seem absurd, and it is hard to imagine that a large part of the population held such beliefs. But our contemporary illusions are not so different. We have rid ourselves of many misbeliefs since the 18th century, mainly thanks to the development of modern sciences, but we certainly did not rid ourselves of all illusions. How would someone from the 23rd century look at our widespread superstitions (touching wood, being apprehensive on Friday the 13th, etc.)? And what would she think of the popular belief that surviving for a week on so-called 'detox' juices and tea clears our body of toxins (which toxins is usually not specified), and that an ethereal, supernatural being was incarnated in a human body some 2000 years ago?

The fact that illusions are part of a coherent worldview and do not look so strange from the inside, only makes it more difficult to expose them. The problem is also that with our intuition or common sense, we can perhaps expose the most outrageous claims, but certainly not all illusions. On the contrary, irrationality often stems from our intuitive and spontaneous thinking. In other words, normal thinking leads us astray. We tumble from one cognitive trap into another. It makes us 'predictably irrational', as the behavioral economist Dan Ariely (2008) describes it.

THREE RULES OF THUMB

Yet we are not powerless. In most cases we can set our thinking straight by using three rules of thumb (Braeckman, 2017). The first rule of thumb is not to accept a claim merely because it sounds plausible. The fact that a claim sounds plausible is absolutely no guarantee that it is true (our intuition can be misleading, as we will find out). So, having the feeling of understanding or knowing something is by no means a guarantee that you actually understand or know something (De Regt & Dooremalen, 2015). This is why we need to rely on external (non-psychological) support to assess its reliability. Is the claim substantiated by facts? Does it emanate from a reliable source?

The amount of external support we should require before accepting a claim depends, of course, on the claim itself. Extraordinary claims must be supported by extraordinarily strong evidence: a photograph of elves or of the monster of Lochness, the 'yeti' or 'bigfoot' are not adequate pieces of evidence for accepting the existence of these creatures. Related to this is the question of the burden of proof. Anyone who comes up with claims about paranormal activities must provide evidence for this, not the other way around Note in this context that despite the large cash prizes promised by skeptics to those who can prove paranormal gifts unambiguously, no one has yet succeeded. Such beliefs are not 'innocent until proven guilty'. The same applies to alternative forms of medicine, conspiracy theories, and other theories that go against the scientific consensus (De Regt & Dooremalen, 2008). Because that consensus is supported by a large amount of evidence and has come about through a reliable process. Anyone who wants to reject the consensus must come with strong counterevidence.

Secondly, we must use "Occam's razor". Occam (an English philosopher from the 14th century) taught us that the most economical or parsimonious explanation is often the best. Such an explanation does not raise many new questions which in turn require an explanation (making the explanation less likely to be true). Take crop circles for example. Some people believe that they are made by extraterrestrials. Another explanation is of course that they are 'hoaxes': that they are made by people to fool other people. Believing the first explanation raises a whole series of other questions that also require an explanation: how did these aliens got here unnoticed, why don't they seek contact, why do they mainly make crop circles in Europe, etc.? The most economical explanation, of course, is that these circles are made by people with a humorous slant.

Finally, we must be aware of a series of 'cognitive pitfalls'. Our thinking is standardly equipped with these pitfalls. Everyone is susceptible to them. In this book we will identify those pitfalls or biases, explain their origin and learn how we can avoid them. Cognitive illusions are similar to perceptual illusions. They are systematic, permanent and universal. Systematic, because our thinking is always distorted in the same way. Human illusions, while they can vary considerably in precise content from culture to culture, are largely variations on the same themes. Cognitive reasoning errors and illusions are also permanent just like perceptual illusions. Take the famous Müller-Lyer illusion:

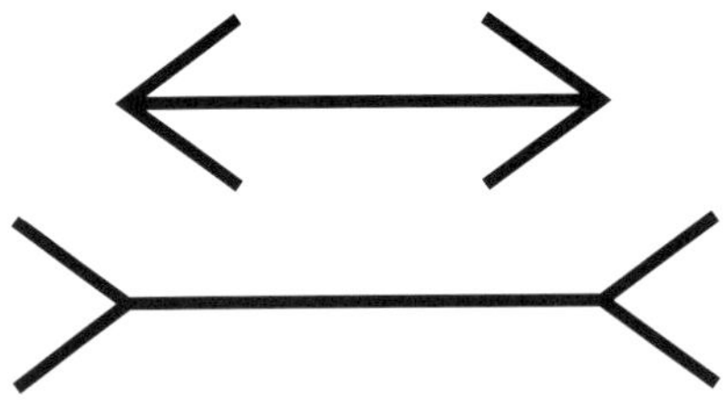

Even if we know that the two lines are the same length (after measuring them for example) and we understand that you are dealing with an illusion, we cannot get rid of the impression that the bottom line is longer than the top line. The same goes for cognitive illusions. Even though we are aware of the cognitive pitfalls (the biases or reasoning errors) that lead to illusions, we still tend to make the same reasoning errors. Finally, illusions are universal. Every normal human brain is susceptible to the same kind of cognitive illusions (just as every human being with normal sensory perception is susceptible to the same perceptual illusions). The first step to critical thinking is therefore to expose the cognitive pitfalls or biases that lead to illusions or irrational beliefs.

PREDICTABLY IRRATIONAL

To learn to think critically, first we need to become aware that our spontaneous thinking is deceiving us in predictable ways. The best way to do this is by exposing you to a series of riddles. They show us in what contexts and in what ways our spontaneous thinking is misleading. For each riddle, before you look at the answer, try to formulate the first answer that comes to mind and then think about why this answer might be wrong.

PROBLEM I: THE 'LINDA' PROBLEM

> Linda is 31 years old, single, outspoken, and very bright. She majored in philosophy. As a student, she was deeply concerned with issues of discrimination and social justice, and also participated in anti-nuclear demonstrations.
>
> Which is more probable?
> 1. Linda is a bank teller.
> 2. Linda is a bank teller and is active in the feminist movement.
>
> (Tversky & Kahneman, 1983)

The answer is 1. If we think logically, we realize that statement 2 cannot be more probable than statement 1, since 2 is a subset of 1. In a study by the psychologists Tversky and Kahneman (1983) it appeared that 85% of the participants answered statement 2. The reason for that is that 2 fits better with the description of Linda, but statistically 2 can never be more probable than 1. Intuitively we are very bad estimating probabilities. That is also evident from the next riddles.

PROBLEM 2: THE BASE RATE FALLACY

> Martin is a single man aged 45. He is an introvert and likes to read. What is more likely: Martin is a librarian (A) or Martin is a salesman (B)?
>
> (Raiffa, 2002, p. 42)

The answer is B and the reason for this is that there are many more salesmen than librarians (approximately a hundred times more). That is why, despite the personality description of Martin, it is still much more likely that he is a salesman. Not taking into account the fact that there are many more salesmen than librarians is known as the 'base rate fallacy' or 'base rate blindness'. The base rate refers to the prior odds or probability. In this example the base rate is the number of librarians in the world divided by the number of salespeople, so 1/100. This number should also be taken into account, not just Martin's personality description.

PROBLEM 3: THE BASE RATE FALLACY

> 1 in 10 000 people suffers from a rare, deadly disease. A doctor develops a test to detect the disease. The test has a false positive rate of 0,5%. So 99,5% of the people who test positive have the disease. Because the test is cheap and very accurate, the government decides to have everyone tested for free. Your test comes back positive. What are the chances that you suffer from the disease? (Kahneman & Tversky, 1985)

Most people answer 99.5%. That is wrong. The figure is much lower. It is only 2% because the base rate should also be factored in: only 1 in 10 000 people suffers from the disease. The chance of a positive test being a false positive (0.5%) is much greater (50 times greater – hence the 2%) than that you belong to the group of people who have the disease (0.01%).

PROBLEM 4: THE BIRTHDAY PROBLEM

> How many people should you gather so that the probability that two people have the same birthday is greater than the probability that no one shares a birthday? [1]

The surprising answer is 23 people, with a group of 57 people the chance has already increased to 99%! Intuitively we think that the number is much higher. (For the statistical calculation – see source above).

PROBLEM 5: EXPONENTIAL REASONING

> Every day a lily pad doubles in size. If after 40 days it covers the entire pond, then at what point would it be covering half of the pond? [2]

39 days (not 20 as we are sometimes inclined to answer immediately, because we reason linearly and not exponentially).

PROBLEM 6: EXPONENTIAL REASONING

> Imagine you could endlessly fold a sheet of paper with a thickness of 0,1 millimeter. How many times would you have to fold it so that the thickness of the sheet reaches the moon (about 385,000 km)? (By folding it once you would get a thickness of 0,2 mm, folding it twice a thickness of 0,4 mm, etc.)

[1] https://www.britannica.com/science/probability-theory/An-alternative-interpretation-of-probability
[2] https://www.riddles.com/1757

# Folds	Thickness (mm)		
0	0.10	22	419,430.40
1	0.20	23	838,860.80
2	0.40	24	1,677,721.6
3	0.80	25	3,355,443.2
4	1.60	26	6,710,886.4
5	3.20	27	13,421,773
6	6.40	28	26,843,546
7	12.80	29	53,687,091
8	25.60	30	107,374,182
9	51.20	31	214,748,365
10	102.40	32	429,496,730
11	204.80	33	858,993,459
12	409.60	34	1,717,986,918
13	819.20	35	3,435,973,837
14	1,638.40	36	6,871,947,674
15	3,276.80	37	13,743,895,347
16	6,553.60	38	27,487,790,694
17	13,107.20	39	54,975,581,389
18	26,214.40	40	109,951,162,778
19	52,428.80	41	219,902,325,555
20	104,857.60	42	439,804,651,110
21	209,715.20		

TABLE 2.1 The exponential growth of folding a piece of paper.

42 times![3] That figure seems absurdly low and that is because we underestimate exponential growth.

Such an exponential reasoning error was often made at the beginning of the Covid-19 pandemic. When the reproduction number was above 1 (meaning that each infected person infects on average more than 1 other person) but there were not that many infections yet, many

[3] https://www.codersrevolution.com/blog/will-a-piece-of-paper-folded-42-times-reach-the-moon

people (including some policymakers) mistakenly thought that the pandemic was under control. But with a reproduction number above 1, you get an exponential growth of the number of infected people. Every so often, the number of infected people doubles. If that is the case – as we have experienced several times – the amount of infected people suddenly increases very rapidly.

PROBLEM 7: THE AVAILABILITY BIAS

> What is more likely: That you will die from a shark attack, or from a dislodged part of an airplane that crashes down? (Tversky & Kahneman, 1973)

In the U.S. the odds are 30 times greater that you die from a broken part of an aircraft than by a shark attack, but since shark attacks get a lot more media attention, we tend to think that they are more likely to happen. This is also known as the 'availability bias'. We often overestimate the likelihood that something will occur when it is easy to recall or imagine (see Appendix).

PROBLEM 8: THE AVAILABILITY BIAS

> Of which are there more English words: Words that start with an R, or words that have an R as the third letter? (Tversky & Kahneman, 1973)

As it turns out, there are many more English words with R as the third letter rather than words starting with R. We tend to think the opposite because it is easier to call to mind words that start with an R than words that have R as the third letter. The availability bias, too, plays a role here.

PROBLEM 9: THE AVAILABILITY BIAS

> What is the probability that a startup will succeed?

It hovers around 10%. We tend to come up with a higher estimate because we hear a lot more about successful startups than failed enterprises. This is known as the 'survival bias' and it is a type of availability bias.[4]

One way in which we express the availability bias is by buying a lottery ticket. Because winners regularly appear in the media, it feels like the chance of winning is real, while it is in fact negligibly small. The probability of winning is around 1 in 8 million. Given that roughly six million people live in Flanders (the Northern part of Belgium), this means that you have a greater chance of bumping into a specific Flemish person whom you don't know, by randomly knocking on a door in Flanders. If people would perceive their chances of winning in this way, the lottery would likely go out of business. Another example of the availability bias (and a similar statistical reasoning error as described in riddle 1) comes from the U.S. Shortly after 9/11, Americans were willing to pay more for a life insurance policy against terrorism than for a life insurance policy that insures against any cause of death.

PROBLEM 10: ANCHORING

> The following 'bias' is known as 'anchoring'. Psychologists (Kahneman, 2011) divided people in two groups. They asked the first group: Do you think the tallest trees in the world (redwoods) are more or less than 300 meters tall? Then they asked them: how tall are the tallest trees according to you? The second group was asked: Do you think that the tallest trees are more or less than 50 meters tall? How tall are the tallest trees according to you?

Interestingly, the answers to the second question varied widely. Group 1 estimated that the tallest trees were 255 meters on average, the average guess of group 2 was 85 meters. The reason for this is that the numbers

[4] https://www.investopedia.com/articles/personal-finance/040915/how-many-startups-fail-and-why.asp.

given in the first question acted as an 'anchor' to the estimates for the second question. Anchoring is a much-discussed bias in behavioral economics and can lead to irrational consumer behavior. We will discover this in chapter 4 ('Irrationality in Action') and in the Appendix ('Detect the Reasoning Errors').

PROBLEM 11: THE FRAMING EFFECT

> The next bias is known as the 'framing effect'. A disease breaks out; 600 people will die if nothing is done. Doctors develop two treatments to combat the disease. Which treatment would you prefer?
>
> Frame A
> *Treatment 1*: 200 people are saved.
> *Treatment 2*: 1/3 chance that they will all be saved, 2/3 that they all die.
>
> Frame B
> *Treatment 1*: 400 people will die.
> *Treatment 2*: 1/3 chance that they will all be saved, 2/3 that they all die.

Test subjects were presented with either frame A or frame B. In frame A, 72% opt for treatment 1, in frame B only 22% and – of course – it comes down to the exact same thing (Kahneman & Tversky, 1979). The decisions we make do not only depend on the information itself, but also to a large extent on the way in which this information is framed (and marketing techniques eagerly play into this as we will see in chapter 4 and in the Appendix).

PROBLEM 12: THE ALLAIS PARADOX

Another type of irrationality in behavioral economics is known as the 'Allais paradox' (Allais, 1953). You have an 80% chance of winning 4000 euro (A) or a 100% chance of winning 3000 euro (B): What would you prefer? Most people choose B (80%). Now, suppose you have a 20% chance of winning 4000 euros (A) or a 25% chance of winning 3000 euros (B): What do you prefer? In this case, most people choose A (65%). However, the consideration in both cases should be the same if we think rationally (in economics this is referred to as 'expected utility theory'). So, we shouldn't shift from B to A. After all, we give up the same sum in B (1000 euros) for the same increase in probability of getting the sum (20% increase).

It becomes even more striking when we make the percentages smaller: suppose you have a 45% chance of winning 6000 euros or a 90% chance of winning 3000 euros, compared to a 0.1% chance of winning 6000 euros or a 0.2% chance of winning 3000 euros. Here, too, we should choose the same in both cases, but intuitively we do not.

With loss, we see the opposite effect. Most prefer an 80% chance of a loss of 4000 euros over a 100% chance of a loss of 3000 euros (92%). But if the percentages change (but not the proportions) we get a different choice. Between a 20% chance of a loss of 4000 and 25% chance of a loss of 3000, 58% opts for the latter.

PROBLEM 13: THE HINDSIGHT BIAS

Another important bias is the 'hindsight bias'. What probability do you think you would have assigned to the breakout of a global pandemic before covid-19 happened? Or of a global financial crisis happening before 2007? Chances are that you might overestimate the likelihood that you would have attributed to these events before they happened, after they have taken place. After the financial crisis, for instance, many investors and economists claimed to have anticipated its occurrence. And yet, most investors still lost quite a bit of money during the crisis, and articles warning about an imminent crisis in 2006 are hard to come by. This is due to the 'hindsight bias'. With 'hindsight knowledge' it often seems quite probable that something would occur. Therefore, after the events took place, we usually think that our predictions in the past were better than they actually were.

The American psychologist Baruch Fischhoff (1975) first identified this fallacy. Before former U.S. President Nixon went to Russia and China, he let test subjects predict the probability of ten possible outcomes of this diplomatic mission. After the presidential visit, Fischhoff asked the participants to remember the probability that they had accorded to each of the ten possible outcomes. Most people, it turns out, overestimated the weight they had accorded to the scenario that unfolded. Most people thought they were better predictors than they really were. This kind of experiment has been repeated several times, for example before and after the trial of O.J. Simpson, and during the impeachment of President Bill Clinton, with similar results.

PROBLEM 14: THE CONFIRMATION BIAS

One of the most truth-distorting fallacies is the confirmation bias. We tend to register and look for information that confirms our beliefs (whilst remaining almost entirely blind to information that contradicts our beliefs).

Psychologist Peter Wason (1960) clearly demonstrated this phenomenon with his 'hypothesis testing' experiment. You are presented with a sequence of three numbers: 2, 4 and 8. A rule corresponds to this sequence. The aim is to guess that rule by proposing different series, each with three digits. Each time you propose a series, you get one of the following answers: 'yes, that follows from the rule' or 'no, that does not follow from the rule'. When you think you know the rule, you can take a guess.

The rule, quite simply, is: a series of increasing numbers. Usually, people take a long time before they find it. The reason that we take a lot of time to solve it, is because we tend to propose sequences that confirm or conform to the rule in our head. We think, for example, that it involves doubling numbers or doubling even digits and propose sequences that correspond to this rule, whilst, in order to solve the riddle, you must do the opposite. You should propose series that do not correspond to your hypothetical rule. That is the only way you can test the rule and, if need be, replace it with another rule, which you can then test again, and so on.

PROBLEM 15: SELF-OVERESTIMATION

Finally, another tenacious bias is self-overestimation. Most of us are susceptible to so-called positive illusions.

85 to 90% of people think that they are above average drivers. 94% of university teachers think that they are above average teachers. 25% of all people think they belong to the top 1% when it comes to social skills. And, ironically, the vast majority of us also thinks that they are less prone to overestimating their abilities than the average person.

Overestimation does not only apply to our sense of self. We also demonstrate a serious bias when estimating the intelligence and talent of our children. We will talk about the evolutionary explanation of these positive illusions in the next chapter. Interestingly, depressed people appear to have a more accurate view of their skills and talents than psychologically healthy people. This phenomenon is called 'depressive realism'.

That concludes the overview of some of our most prominent biases. As Dan Ariely (2008) summarizes it: 'Even the most analytical thinkers are predictably irrational; the really smart ones acknowledge and address their irrationalities'. That is the purpose of this book.

SUMMARY

What is critical thinking?

Rational and autonomous thinking

What are the three rules of thumb of critical thinking?

1. Demand external (not psychological) support for beliefs.
2. Apply Occam's razor: choose the most economical / parsimonious explanation.
3. Beware of cognitive illusions.

Why Are We Irrational?

The evolutionary origin of irrationality

THE PECULIAR ARCHITECT OF OUR THINKING

In the previous chapter you experienced how your thinking is predictably irrational in certain contexts. In this chapter you will discover why this is the case. To answer this question, we must first turn to the architect of our thinking. Just as we would turn to the technicians of, say, a calculator, if it turned out that the calculator performs certain calculations incorrectly. Since Darwin, the architect of our thinking apparatus is known: evolution by natural selection. This remarkable architect appears to be blind (i.e. without foresight or plan), has no say in the materials with which it works, and has only one goal: reproduction. This has a series of important consequences for our thinking. Before outlining these consequences, it is useful to briefly describe the process of evolution by natural selection.

EVOLUTION BY NATURAL SELECTION

Evolution - the fact that species change over time and that all life forms (on our planet at least) share a common ancestor - is mainly driven by natural selection. Other factors influencing evolution are sexual selection, and, to a lesser extent, genetic drift and epigenetics. For the purpose of this book, we can limit ourselves to evolution by natural selection. That process consists of three steps. Firstly, there are random genetic mutations (copying errors in the DNA of an organism). This creates genetic variation (genetic differences between the individual organisms in a population). Secondly, those genetic mutations are passed on to offspring. Thirdly, mutations that help the organism survive and reproduce in its environment are 'selected'. Because organisms with those mutations have a greater chance of surviving and reproducing, more organisms with these genetic mutations will be present in future generations ().

Take, for example, the long neck of a giraffe. That neck has grown over time because giraffes with a genetic mutation for a slightly longer neck reproduced more than giraffes with a shorter neck, since the animals with a longer neck were better able to feed on tall trees. Throughout generations, that neck continued to grow, because in every generation the animals who happen to have longest neck got the most food and therefore had the greatest chance to breed and to pass on their genetic material which coded for a long neck. Mind you, natural selection is blind: it has no plan (such as making a long neck for giraffes). Every generation, organisms with different traits had different success in reproducing and so, over time, 'adaptations' to the environment gradually emerge (the entire population possessing beneficial traits). Moreover, the blind architect (natural selection) has no control over the material it works with, since the mutations that arise are random (the majority of these mutations are neutral or detrimental to the organism and are therefore not selected).

WHAT DOES THIS ENTAIL FOR OUR THINKING?

Undeniably, our brains are also the product of natural selection. Because natural selection cannot anticipate the future and does not always have the optimal mutations to select from, the process often results in suboptimal designs. Take our eyes, for example. They have evolved from light-detecting cells under the skin. The nerve-bundles came together at the top in these cells. When those cells gradually evolved into the complex eye with pupil and retina, those nerves remained at the front of the eye, which means they had to be drilled through the retina to connect to the brain. That is why we have a blind spot in the visual field of each eye (a problem that is solved by combining the field of vision of the two eyes). Squid eyes have evolved separately (our common ancestor with squids had no eyes) and are better designed. In squids the nerves come together behind the eye, so they have no blind spots.

So, with regards to our brains, we can assume that more optimal designs are possible, at least in principle. But more importantly, the reason for our thinking errors, is the 'goal' of natural selection. Natural selection is only 'interested' in reproduction. It drives the evolution of a species by the different success with which genes (genetic variations) spread in the population. In a way, genes use organisms as vehicles to make copies of themselves (by enabling these organisms to reproduce). They are successful to the extent that they provide the organism with characteristics that increase its chances to reproduce. For example, by providing the organism with adaptations that make it better able to survive (such as camouflage, sharp teeth or long necks) or that make it more attractive to the opposite sex (such as the colorful tail of the peacock - that is sexual selection).

TRUTH IS AN EXPENSIVE MEANS TO AN END

Each characteristic of an organism, therefore, is selected only insofar as it yields a reproductive advantage. The same goes for our brains. They have not evolved to provide us with true representations of the world, but with representations that increase our chances of survival and reproduction. Truth (representing the environment in a correct way), however, is usually the best strategy to increase the chances of survival and reproduction of an organism. Take two hominids who see three tigers enter a cave and see two tigers come out. The hominid who made the right calculation and deduced that there still was a tiger in the cave, is more likely to be our ancestor (and to have passed on his mathematical genes) than the one who thought the coast was clear and moved in.

But truth comes at a cost. Representing the world in a complex and accurate way requires a lot of brain power. This, in turn, requires a lot of food. Brains are expensive organs. Our brains exhaust 20% of the energy we get out of food, while they only make up 1-2% of our body mass. More brain power (and brain mass to sustain it) can only evolve if the benefits it generates for the organism (in terms of survival and reproduction) are greater than the additional cost it requires (the extra food that must be found). So, natural selection is interested in truth only to the extent that it is relevant for survival and reproduction and wants that truth to be as inexpensive as possible. This has a series of important consequences.

SYSTEM 1 AND SYSTEM 2

First and foremost, our thinking apparatus was developed to function rapidly and economically. Complex thinking processes require a lot of time and energy. The hominids in our example above did not have the luxury to think for a long time about if there actually was a tiger in the cave. Nor did they have the luxury to engage in overly complex forms of information processing, because that would require a brain that is (even) more costly and they would need (even) more food to sustain it.

As a result, we are equipped with a thinking system that is both fast and frugal (economical). It works automatically, quickly and intuitively. Cognitive psychologist Daniel Kahneman (2011) calls this cognitive mechanism 'system 1'. We do, however, also have a second system at our disposal; a system that can check the output of system 1 and overwrite it, if necessary. This 'system 2' is slow, conscious, and requires effort. In general, system 1 is in control. When system 1 is in control, our thinking operates on 'automatic pilot'. To have you experience the difference between system 1 and system 2 thinking (and discover which system is in control), let me present you with two additional riddles.

The first riddle is commonly known as the Moses illusion. It goes like this: "How many animals of each kind did Moses take in the Ark?" (Erickson & Mattson, 1981).

The answer, of course, is that Noah, and not Moses, took animals on his ark. But system 1 is inclined to answer "two" immediately (it works fast and automatically), because Moses fits in the biblical context.

The second riddle yields a similar result. "A baseball bat and a ball cost 1 $ and 10 cents together. The baseball bat costs 1 $ more than the ball. How much does the ball cost?" (Kahneman & Frederick, 2002).

Here, too, system 1 is inclined to answer ten cents almost immediately. If we think it through (with system 2), however, we see that the correct answer is 5 cents: the bat costs one dollar and five cents and the ball five cents, which adds up to one dollar and ten cents.

THE FALLIBILITY OF SYSTEM I

Heuristics: simplicity trumps complexity

So, system 1 regularly leads us astray (remember the riddles in the previous chapter!). The reason for this is that it makes use of heuristics. Heuristics are simple rules of thumb that generally produce good results but can sometimes be misleading. They are natural selection's solution to finding as much relevant truth about the environment in the most 'cost-effective' manner. System 1 is a system of approximation: is applies simple rules to get as much relevant truth as possible as quickly and cheaply as possible. Although reasonably effective, it is therefore also fallible.

The application of heuristics gives rise, for example, to the availability bias (e.g. the fact that we are inclined to think that deaths from shark attacks are more common than deaths caused by dislodged aircraft components, since the former is easier to imagine or recall than the latter – see previous chapter – problem 7). In this case, the heuristic we apply unconsciously and automatically is: 'The easier it is to recall or imagine an event, the more likely that event is'. This is often true, but not always.

Error management

As such, system 1 can be misleading, because it replaces complex reasoning processes that require a lot of information (e.g. statistical calculations) with simple reasoning processes that can be carried out in a split second. It does so to enable us to come to conclusions in a quick and economical way. In some cases, however, system 1 has been manufactured by natural selection in a 'deliberately' misleading way. Remember that the 'purpose' of natural selection is not truth, but survival and reproduction of the organism. Since some mistakes are more costly than other mistakes (i.e. more threatening to survival and reproduction), natural selection will primarily attempt to avoid these costly mistakes. To this end, it is willing to make more mistakes.

Compare it to a fire alarm. The alarm can make two errors: it can go off and signal that there is a fire when there is no fire (a false positive) or fail to go off when there actually is a fire (a false negative). The second error obviously has more serious consequences than the first one. That is why fire alarms are designed in such a way that they go off too easily (e.g. when smoking a cigar in the room) to make absolutely sure that they will not generate a false negative (failing to go off when there is a fire). The goal of the maker of the fire alarm is not to keep the total number of errors (both false positives and negatives) as low as possible, but to keep the cost of errors as low as possible. So, to avoid the costly false negatives, it yields quite a few false positives (and more errors in total).

The same goes for system 1. Natural selection has not designed it to be as accurate as possible, but to avoid costly mistakes. For that purpose, it is prepared to make a larger total number of mistakes. This phenomenon is known as 'error management' (). A good example of error management is the tendency of men to overestimate the interest women have in them. Evolutionary, this makes sense. Since the cost of a missed opportunity for reproduction is much higher than the cost of a fruitless attempt to seduce, natural selection 'wants' to avoid errors of the first type and is therefore prepared to make more errors of the second type (Haselton & Buss, 2000).

Another example is detecting causal connections. Much like a fire alarm, this detection mechanism is tweaked in such a way as to make sure that we will not miss any important causal relations (such as the relation between eating something toxic and getting sick) because in general missing causal relations is more costly than seeing causal relations that are not there. The result is that we often tend to see causal relations that are not there. This, in turn, is a breeding ground for superstition, pseudoscience, and conspiracy theories. We will discuss this is more detail in the next chapter.

Finally, a third example of 'error management' is something commonly referred to as 'hyperactive agency detection' (Barrett, 2000). We tend to discern the actions of an 'agent' (a living being with intentions) in certain events too quickly. The classic example goes as follows: a hominid sees a bush move and hears it rustle. It can either be the wind or a predator stalking in the bushes. Thinking that it is the wind while it is a predator (a false negative) is a lot more costly than making the opposite mistake (a false positive). So, again, we are inclined to make too many false positive mistakes to avoid the costly false negative mistakes. This 'hyperactive agency detection' plays a particularly important role in the emergence of supernatural (religious) beliefs – as we will see in the third chapter.

Evolutionary mismatch

In addition to the fact that system 1 is an 'approximating' system (because it is economical) and that it sometimes makes a larger total number of mistakes to avoid costly mistakes (error management), there is a third reason why system 1 sometimes deceives us. The heuristics that make up system 1 have been designed to guide us through the environment in which we have spent most of our evolutionary history, not in the environment in which we live today. Our thinking is adapted to a nomadic existence in the Stone Age (since we spent the vast majority of our evolutionary history as nomadic hunter-gatherers). Sometimes our 'stone age minds' yield bad results in the modern world (Tooby & Cosmides, 1992).

A good example of this is the 'gambler fallacy'. When we toss a coin, for instance, we tend to think that the probability of getting a 'head' increases the more 'tails' have been tossed consecutively. In other words, we tend to expect a statistical correction. Of course, the chance is always 50/50, regardless of what came up in the past. The same applies to the roulette wheel. Gamblers tend to play red when the ball ends up in a black pocket a few times in a row, 'because 5 times black in a row, would be really unlikely'.

In a casino, this is obviously irrational. But in the natural environment in which we have evolved, this way of thinking makes sense. After all, most natural events are cyclical. When predicting the weather, for example, the probability of rain does increase as the length of a dry period increases (in a climate and season in which it rains regularly). In other words, it is mostly in artificial, modern contexts (the casino) that this way of thinking is irrational (Pinker, 1997). The same can be said for the many statistical reasoning errors we tend to make (like the base rate fallacy – see examples in chapter 2 and in the Appendix). We did not encounter these problems in our ancestral environment, so our intuitions did not evolve to solve them.

TAKING STOCK

So, let us take stock. Evolution has provided us with two cognitive systems: the first is fast and frugal. It leads to reasoning errors because of its approximating nature, error management and a 'mismatch' between the problems it was designed to solve (stone age problems) and the problems we face today. System 2 can put our thinking back on track, but it requires effort, and it usually stays in the background. (Kahneman, 2011) calls system 2 'the lazy controller'. System 2 only intervenes if there is no response from system 1 (for example when you calculate 25x56 – since we do not have an immediate, intuitive answer for this) and when we deliberately switch it on to check the output of system 1 (by consciously reflecting on a problem).

Now, this is precisely what the critical thinker does: they recognize the fact that their automatic, intuitive thinking is fallible in certain contexts and call upon system 2 (our conscious, reflective thinking) to check the output of system 1 in those contexts. Because system 1 can never be switched off, those answers come automatically. It is therefore a matter of checking the constant flow of system 1 when called for, and not following it blindly.

OTHER SOURCES OF IRRATIONALITY

The social environment

So far, we have only discussed the evolutionary need to navigate our physical or natural environment. But humans are also part of a social environment. As social primates, our survival and reproductive opportunities depend to a large extent on our relationship with other members of the group. Our cognitive apparatus is therefore not only designed to navigate the physical-natural environment, but also the social environment. This has some important consequences. Bear in mind here that our thinking has not evolved for the purpose of truth but for reproduction. Truth is a means to an end. But, as I have previously pointed out, to successfully reproduce, it is usually best to represent the natural environment truthfully. However, that only applies to the natural environment. In the social environment, truth is a lot less important. In fact, in this context we often benefit from deceiving others (as long as they do not detect it, that is).

In a conflict, for instance, I benefit from the fact that my opponent believes I am a greater threat than I actually am (e.g. by overestimating my physical strength or the number of people in a group who would take my side). This increases the chance that the opponent withdraws, so I get what I want, without having to engage in a potentially costly fight. The same goes for my status in the group. My talents are better overestimated than underestimated. That way I benefit from a higher social status and see my chances of survival increase as well as my reproductive opportunities. To deceive others successfully, natural selection has equipped us with a resourceful 'bias'. The best way to deceive others is to deceive yourself. The liar who does not know that they are lying, is often the best liar. This explains the overestimation of one's own talents and prospects that we discussed in the previous chapter (problem 15). Here, too, our spontaneous, intuitive thinking system – system 1 – leads us astray.

The irrationality of system 2

But our irrationality is not solely attributable to system 1. Our slow and conscious thinking processes (system 2) also systematically deceive us in certain ways. This is due to its function in the social environment. Reasoning is not only used to obtain relevant insights about our environment (i.e.to navigate the environment in a way that contributes to survival and reproduction), it is primarily (!) used to argue with others. When arguing with others it is not as important to be right as it is to persuade others that you are right. As such, natural selection has equipped us with reasoning capabilities that are designed to persuade, since people who were able to persuade others and win arguments saw their social status rise, and with it, their chances of survival and reproduction. Cognitive scientists Dan Sperber and Mercier (2017) draw this very conclusion in their 'argumentative theory of reasoning': reasoning evolved for arguing.

As a result, we are by nature lawyers (great at arguing and persuading), rather than philosophers or scientists (not so great at finding truth by questioning our own opinions). The cognitive mechanism that helps us with persuading others that we are right, is the infamous 'confirmation bias'. It helps us defend our opinions and convince others that we are right, by making us see and remember the evidence and arguments supporting our opinions (and filtering out the counterevidence and arguments). This often comes at the expense of being right.

If a prize were to be awarded to the bias that distorts our thinking most profoundly and regularly, it would undoubtedly be granted to the confirmation bias. The confirmation bias selectively suppresses everything that could contradict our beliefs and opinions. We are blind to counterarguments and see confirmation for our own convictions all around us. It predisposes us to seek, observe, remember, and interpret information in such a way that it reinforces our pre-established points of view. Because the confirmation bias makes us see and retain confirming information, whilst filtering out the counterevidence, it creates another bias: the overconfidence bias. This refers to our own tendency to (grossly) overestimate the odds that we are right (since we overlook and quickly forget the counterevidence).

Interesting research conducted in the 1970s clearly reveals the workings of the confirmation bias (Lord et al., 1979). Psychologists presented several studies on the relationship between the death penalty and crime rates to a group of people. Half of the group supported the death penalty, the other half was opposed to it. The studies – which were fictional, but the participants were not aware of this – contradicted each other. Some studies concluded that crime was decreasing with the introduction of the death penalty, whilst other studies indicated that there was no link between crime rates and the death penalty. One would expect that most of the participants would take a more nuanced stance regarding the issue (since no univocal conclusion emerges from the studies). However, the opposite turned out to be the case. Not only did all of the participants stick to their points of view, they did so with (even) more conviction!

Participants invariably estimated that the studies supporting their position (for proponents of the death penalty: a negative correlation between introducing death penalty and crime, for opponents a lack of such a correlation) were better than the studies contradicting their views (they only saw flaws in the latter and interpreted ambiguous information in favor of their opinion – this is known as the 'belief bias', it's described in the Appendix). More strikingly, they seemed to recall the studies supporting their point of view much better than the studies contradicting it. This kind of research also shows that the confirmation bias gains force when one is emotionally involved in a debate.

Emotion

This brings us to another prominent distorter of truth: emotion. We are not dispassionate robots objectively analyzing the world, but hotheaded primates who see the world through a prism of emotions. The reason that natural selection has provided us with emotions in addition to information processing is simple: natural selection is only interested in actions (that are conducive to survival and reproduction). And in order to induce an organism to act you need two things: a belief (the result of information processing) and a desire (the result of emotions or feelings). For example, I am encouraged to open the refrigerator and get something out of it (action) because I know that there is food in there (information processing) and because I am hungry (feeling). The affective component is the driving factor of action, and the cognitive component (information processing) is the guiding factor. But the two components are not separate from each other. The affective, in particular, influences the cognitive. This is known as the 'affect heuristic'.

The affect heuristic consists of making decisions (e.g. whether or not to make an investment or get an insurance) on the basis of emotional reactions, not on the basis of the information available and an objective cost-benefit analysis (remember the Americans who were willing to pay more for a life insurance policy against terrorism than a life insurance policy where every cause of death is covered in chapter 2 - problem 9). In weighing risks against potential return, we often let ourselves be guided by our 'gut feeling', not by objective information and rational analysis.

The consequences of this can be dramatic. In addition to making us bad investors, it also makes us bad policy makers. With regards to climate change, for example, the effects of the affect heuristic turn out to be disastrous. Most of us are aware of climate change and yet we are doing relatively little to control it. One important reason for that is that the affective response we have in relation to climate change is rather small. Compare this with terrorism, for example, which provokes a much stronger affective response but, objectively, is a much less serious threat.

A particularly strong affective disposition with which we are all equipped is the so-called 'ingroup - outgroup bias'. We have a positive disposition towards members of the group to which we ourselves belong (ingroup) and a negative disposition toward people from other groups (outgroup). This has profound negative consequences for human society (such as war and racism) and it also distorts our thinking.

We tend to trust sources within our group too easily and therefore take over irrational beliefs form the ingroup, while we are typically extremely skeptical of the beliefs of members of the outgroup. This also plays an important role in the spreading of religious beliefs. Moreover, it explains why we tend to take on the irrational beliefs of our own group without a second thought, while the irrational beliefs of other groups often seem completely absurd (see chapter 1).

Together with the confirmation bias, the ingroup - outgroup bias is one of the biggest stumbling blocks to critical thinking. In chapter 5, we will look at how we can protect our thinking from these pervasive biases. In the next chapter, we will look at the various domains of irrationality that these biases lead to.

SUMMARY

What are our two thinking systems?

System 1: fast, automatic, intuition-based system

System 2: slow, effortful, reflection-based system

Why is system 1 fallible?

- It is frugal: it chooses simplicity above complexity
- Error management
- Evolutionary mismatch

Why is system 2 fallible?

Adapted to the social context: designed to convince others, win arguments

Which reasoning error helps with this?

The confirmation bias

What is the third source of irrationality?

Emotions

How does evolution by natural selection work?

1. Variation: random genetic mutations occur.
2. Replication: those genetic mutations are passed on to the offspring.
3. Selection: genetic material coding for traits that help the organism survive and reproduce will be more prevalent in subsequent generations. Since organisms possessing those traits will on average be more successful at reproducing (and passing on the genes coding for these traits) than organisms not possessing those traits.

FURTHER READING

Kahneman, D. (2011). *Thinking, fast and slow.* Farrar, Straus and Giroux

Sperber, D., & Mercier, H. (2017). *The enigma of reason. A new theory of human understanding.* Allen Lane & Harvard University Press

Irrationality in Action

How reasoning errors lead to domains of irrationality

4

In the previous two chapters (and in the Appendix), we saw how our spontaneous thinking deceives us as well as why that is the case. On the one hand, we have our automatic, intuitive thinking (system 1) that works well in most cases but can be deceptive because it is fast and frugal (economical). Moreover, we saw that system 1 sometimes 'deliberately' misleads us in order to avoid making costly mistakes (error management) and misfires in certain modern contexts (mismatch). On the other hand, we have our slow, reflective and conscious thinking (system 2) that helps us critically analyze the output of system 1 and overwrite it if necessary. System 2, however, is not infallible; it subjects us to a strong confirmation bias - the tendency to be receptive only to confirming information and evidence. Finally, our emotions must also be considered. Our thinking can be affected by our feelings. A prominent bias that arises from this is the so-called 'ingroup - outgroup bias': we take on beliefs from within our group rather easily (and mistrust sources outside of the group). These systematic and universal 'biases' underlie domains of irrationality or illusions: sets of commonly held beliefs that misrepresent reality. In this chapter, we look more closely at these domains.

SUPERSTITION, HOROSCOPES, AND PALM READING

An important source of irrationality is our 'over perception' of causal relations. As explained in the previous chapter, it was generally less costly of a mistake for our ancestors to see causal relations that were not actually there, than the mistake of failing to detect causal relations that were there. To that end, natural selection has equipped us with a mind that is prone to over-detect causal relations (to make sure we do not overlook any).

This bias leads to superstition. Superstition arises because we think we perceive all kinds of patterns and relations that are not really there. Horoscopes establish links between the positions of celestial bodies and our personal experiences here on earth, and psychics draw relations between the lines in the palm of our hand and our future. Once we have become receptive to these false relations, our confirmation bias reinforces the belief that these forms of divination are truly onto something, because it makes us much more receptive to confirming evidence than to disconfirming evidence.

On a more basic level, we are all prone to make such false connections. Think of the athlete who wears the same pair of 'lucky' socks at every competition, or the apprehension of some people to walk under a ladder or on Friday the 13th, or the fact that they keep a talisman for good luck. They all see a relation (between an object or number and luck or unluck) that is not there. Superstition is as old as humankind itself and will probably exist for as long as humans do. It is an (unintended) by-product of the fact that the architect of our mental apparatus (natural selection) wanted to make sure that we did not overlook any causal relations in our environment.

CORRELATION DOES NOT IMPLY CAUSATION

Even if there actually is a correlation between two events, that correlation – contrary to what we tend to think – is not always causal. Correlation does not imply causation! A good illustration of this occurred at an Israeli air-force base. The instructor noticed the following: when congratulating a pilot in training after a well-executed exercise, that pilot often performed worse in their next exercise. Conversely, when the instructor reprimanded the pilot for poor execution of said exercise, that pilot's performance often improved in subsequent attempts. The instructor deduced that punishment works better than reward and that from now on only punishment should be used.

The reason for the correlation, however, had nothing to do with the instructor's praise or scolding, but simply with the fact that after an exceptionally good performance, statistically, the chances of a weaker performance are much higher than those of a similar or better performance. Conversely, for a weak performance, statistically it is more likely that it will be followed by a better performance. This phenomenon is called 'regression to the mean,' and is often overlooked in alternative medical circles (see below).

ORDER IN RANDOMNESS AND RANDOMNESS IN ORDER

Since we are inclined to detect causes and patterns too quickly, we often underestimate the role of chance, randomness and coincidence. People see patterns in random sequences (e.g. in random sequences of numbers) and unconsciously put patterns in sequences of numbers when they are asked to make them random (because they put too much alternation in the sequences, such as never repeating the same number in a sequence or never having three or more heads or tails in a row in a sequence of random coin tosses – whereas in actual random series that often happens). So, we see order in randomness and imitate randomness with order! Noteworthy in this context is that the first iPod Shuffle which played songs randomly was quickly adapted because there were many complaints from people who were convinced that the succession of songs was not random. The manufacturers ended up putting patterns in the sequence of songs to make it seem random!

CHANCE BLINDNESS

The chance factor is also grossly underestimated. Athletes who score a few points in a row are said to be on a 'confidence streak,' whilst their colleagues who record a few misses in succession are thought to have 'broken under the pressure'. While psychology plays a role in athletic performance and these analyses could very well be true, statistical analysis in basketball has shown that these alleged clusters of hits and misses turn out to be random. Nevertheless, matches are still analyzed in such a way because it seems that there is indeed a connection between a player's previous hits or misses and the probability that he will score the next time. For this phenomenon, psychologists came up with the term 'hot hand fallacy' (Gilovich et al., 1985). This bias does not only affect sports commentators and coaches; we all tend to underestimate the importance of chance. Investment success is often wrongly attributed to insight and CEOs and coaches are often rewarded after successes or fired after disappointing performances, while both success and failure are, to an important extent, the products of unpredictable external factors.

Finally, when we are confronted with extreme coincidences, we often refuse to see those events as merely accidental. Not only is it a possibility that such coincidences occur, it is a statistical certainty that they will occur regularly. The chance that you win the lottery may be very small, but the chance that someone wins the lottery is not. Examples of such extreme coincidences abound: the house of the French family 'Comette' was destroyed by a comet, James Dean's Porsche repeatedly brought misfortune to people (look it up!), a baby was saved twice when falling out of a window by the same man who walked under it, and 70-year-old twin brothers both died in a car accident on the same road in Finland in a timespan of two hours. In such cases we often tend to think that 'this cannot possibly be a coincidence', but of course it is.

CAUSAL REASONING ERRORS

As cause-thirsty beings, we also routinely make up causes for events when we do not know their real causes. Comets and solar eclipses were often seen as signs of the wrath of the gods before we had scientific explanations for these phenomena. In short, we are overzealous in finding causal relations: we see many causal relations that are not there, and if we cannot find out the real causes, we just make something up.

Moreover, we also misinterpret causal relations. A common fallacy is that we confuse the probability that A if B with the probability that B if A. We jump to conclusions and think, for example, that someone is angry with us when that person does not answer our telephone call. But the chance that said person would not answer their phone if they were angry with us has nothing to do with the probability that that person is angry with us when they are not answering the phone. There are many other explanations for this: they might have left their phone somewhere, or are too busy to respond, etc. Yet, we often fail to consider these other possible causes. This kind of causal fallacy also plays an important role in conspiracy theories. Such theories are especially interesting because they tend to be fueled by a series of different biases and cognitive illusions.

CONSPIRACY THEORIES

Conspiracy theories – such as the theory that the moon landing in 1969 never took place, that both former U.S. president John F. Kennedy and the British Princess Diana were assassinated by their governments, that Georges W. Bush orchestrated the 9/11 terrorist attacks, or that covid vaccines contain microchips to monitor the population – provide accounts of events that, to say the least, differ strongly from their 'official' version. The official version is seen as a cover story set up by the guilty party. Conspiracy theories often arise from basic causal reasoning errors. For example, given that the terrorist attacks of 9/11 helped increase public support for the Bush administration and, to a certain extent, enabled former president Bush to invade Iraq in 2003, the Bush administration is thought to have orchestrated the attacks. But, of course, it is not because 9/11 was used for political purposes and increased support for the Bush government, that it was also staged by said government (Braeckman & Boudry, 2011).

The ingredients for conspiracy theories

Conspiracy theories are not fueled solely by causal reasoning errors. As these theories develop, a whole series of other biases come into play. Unsurprisingly, the confirmation bias is most obviously at play. Conspiracy theorists focus almost exclusively on 'evidence' that supports their theory (for example, a single eyewitness who claims to have observed something remarkable) and mainly surround themselves with like-minded people (in the US large gatherings take place of the so-called '9/11 Truth movement'). The latter plays into the so-called ingroup - outgroup bias. Like-minded people belong to the ingroup in which everything they claim is accepted uncritically. People who reject these theories are part of the outgroup and their arguments often fall on deaf ears. Sometimes they are even accused of being part of the conspiracy!

In this way, conspiracy thinkers shield their theory from any criticism. From the view of the conspiracy theorist, hard criticism can be expected because the opposing party naturally wants to suppress the truth, which conspiracy theorists think they have discovered. Strong and pointed criticism is therefore not seen as casting doubt on the theory, but typically interpreted by conspiracy theorists as an indication that they are on the right track! In this sense, conspiracy theories possess a kind of built-in immunization strategy (a feature that protects the theory against refutation). The more one points out flaws in these theories, the more adherents see it as a sign that they are right (Braeckman & Boudry, 2011). Shielding a theory from criticism is a core tenet of uncritical thinking. Such immunization strategies also play a major role in pseudoscience (as we will see below).

Debunking conspiracy theories

Conspiracy theories are usually quite easy to debunk with a healthy dose of critical thinking. In general, conspiracy theories tend to become increasingly less likely as they evolve. The reason for this is that advocates of conspiracy theories make statements which require explanations and as such force them to come up with supporting evidence and arguments that tend to become progressively more farfetched.

Take the 9/11 conspiracy theory, for example. One of the most important arguments invoked by its advocates, is that the WTC towers collapse too quickly and in too 'controlled' a manner. Conspiracy theorists think that the towers were brought to explode with bombs placed on the inside by the Bush government and were detonated shortly after the aircrafts flew into the towers. But that raises a whole series of subsequent questions: who put those bombs there? Probably not President Bush himself. If not him, then perhaps a team sworn to secrecy? But how many people are needed for this? To execute such an operation, at least a few hundred people must be part of the conspiracy. All of them Americans, many of whom would have known people in New York. Why did nobody disclose information about the imminent attacks before it happened to warn friends and family? Why would no one have confessed these plans simply out of conscientious objection? All of this seems quite unlikely, especially in times of whistleblowers.

When faced with such theories, we should remember the second rule of thumb of critical thinking, Occam's razor: the most economical explanation is usually the most probable one. The more questions a theory raises (all of which in turn require further explanation), the less likely the theory itself becomes. These conspiracy theories often remain rampant because – like pseudosciences – they typically lack an internal, self-critical feedback loop.

PSEUDOSCIENCE

This, then, brings us to the question of what distinguishes science from pseudoscience. Philosophers have been reflecting on this issue since antiquity. Aristotle thought that science produces knowledge about the true causes of phenomena, whilst pseudoscience only conveys unfounded opinions. But how can we know whether we are actually dealing with scientific theories? Aristotle's criterion: 'knowledge of true causes', appears both too strong: good scientific research often turns out to be (partly) wrong or at least incomplete (think of Newton's physics, for example) and useless: we can only determine with hindsight that a theory was wrong.

Popper's demarcation criterion

Reflecting upon this central issue in the philosophy of science, the 20th century philosopher, Karl Popper (1963), proposed a 'demarcation criterion' that remains very influential to this day. According to Popper, a theory is scientific if, and only if, it can, in principle, be refuted (proven wrong) by means of observation. This may sound rather odd. Is it not the likeliness of being true rather than the possibility of being false that makes a theory scientific? Popper turns our intuition about reliable knowledge on its head. It is not certainty, but fallibility that defines science. And there is a good reason for that.

As we saw in the previous chapter, the bias that deceives us most frequently is the confirmation bias. We are primarily focused on evidence that supports our own theory and are selectively blind towards evidence that would refute our theory or belief. A good thinking system, then, is a system that protects us against the confirmation bias. From a Popperian perspective, it is exactly this protection against the confirmation bias that should form the backbone of scientific thinking.

Freud vs Einstein

This insight came to Popper (1963) while he was reflecting upon two very influential theories in the first half of the 20th century: Sigmund Freud's psychoanalysis and Albert Einstein's theory of relativity. Both theories revolutionized their respective scientific fields: Freud introduced the workings of the subconscious mind in psychology, and Einstein introduced the relativity of time and space in physics. Both theories also had a wide-ranging explanatory scope: Freud could explain many psychopathological conditions by invoking subconscious psychological dynamics, and Einstein's theory could explain everything Newton's theory explained and more, given that it explained the anomalies in Newton's system, such as the orbit of Mercury around the sun.

Unsurprisingly, both theories were seen by many as very successful. But there was one crucial difference between the two theories according to Popper: Freud's theory could not be debunked by observable facts, whereas Einstein's theory could. For Freud, all possible instances of human behavior could be explained from the same set of principles.

When philosopher of science Sidney Hook (1959) asked an auditorium full of psychoanalysts what kind of behavior a child should exhibit for it not to suffer from the Oedipus complex, the room kept remarkably quiet. In other words, no conceivable kind of behavior could refute Freud's pet theory.

Einstein's theory, on the other hand, makes precise predictions that are testable, and has undergone and passed numerous tests. During a solar eclipse, astronomer Arthur Eddington observed that starlight was deflected by the mass of the sun, just like Einstein had predicted. An atomic clock was brought into orbit around the earth at high speed, and it showed a slight distortion of time compared to the time measured on earth, as Einstein had predicted. Einstein's theory is currently still being tested, for example at the CERN facilities. If it turned out that a particle in the accelerator would reach a speed faster than the speed of light, Einstein's theory would be falsified.

The importance of criticism

That is the crucial difference between science and pseudoscience. Science remain open to refutation, whereas pseudoscience protects or immunizes its theories from refutation. Certainly, that does not mean that scientists are not susceptible to the confirmation bias. Of course they are – we all are – and good scientists are aware of this. Charles Darwin, for example, made the following entry in his journal: 'I used the following golden rule for many years: as soon as I noticed a published fact, a new observation or thought that contradicted my general results, I always made a note of it. I knew from experience that such facts are very easily overlooked and forgotten.' (Darwin, 1887/1958).

However, we cannot expect all scientists to have so much insight into their own psychology and to be as diligent as Darwin. Fortunately, the scientific context and methodology protects the sciences from the scientists' confirmation bias. By creating an environment of open debate which encourages participants to approach each other's work critically, the truth-biasing effects of the confirmation bias can be kept at bay. In such a context, the confirmation bias can even be an asset. This way, scientists go the extra mile to defend their hypotheses, while their colleagues go the extra mile to come up with counterevidence and counter arguments. In the scientific context, there is no lack of motivation to refute a theory (Boudry, 2016). The physicist who refutes Einstein's theory goes in the history books.

Immunization strategies

This environment of open criticism is typically not present in pseudosciences. In pseudoscientific circles, discussion with skeptical colleagues is not prevalent. Theories are typically not made vulnerable by exposing them to criticism but are shielded from criticism. This is done by means of so-called immunization strategies. Firstly, pseudoscientists often weaken their claims or give a new interpretation when there is strong counterevidence. In other words, they set up 'moving targets'. A good example of such a re-interpretation occurred in the religious community known as the witnesses of Jehovah. They predicted that Christ would return in 1873. When he did not, they argued that he had indeed returned but as an invisible spiritual being.

We find a similar move in Freud's psychoanalysis. According to Freud, neurotic disorders are the result of a frustration of the libido. When many soldiers developed neuroses during the first World War by being exposed to the horror of war (the shellshock phenomenon currently referred to as 'post-traumatic stress disorder'), refuting Freud's view, Freud saved his theory by arguing that war threatened the soldiers in their most desired love object: their own body. The libido theory remained intact, but it was radically reinterpreted.

Another technique frequently used in pseudosciences to protect a theory from refutation is to build in enough vagueness in the theory. Chakra stimulation, for example, consists in cleaning 'chakras' from 'bad energy' and, doing so, in healing the patient from their ailments. If there is no immediate improvement, the therapist can easily explain that away by saying that the chakras are more thoroughly blocked than initially thought. Given that many ailments heal spontaneously over time, there is usually some improvement in the end, after which the therapist gets all the credit.

Healthcare: a perfect storm!

This brings us to the domain where pseudoscientific theories tend to be most prominent: that domain is healthcare. This is no coincidence, a combination of factors creates 'a perfect storm' when it comes to healthcare. Three factors in particular: first, there is the confirmation bias of the therapist, second, the placebo effect on the patient, and third, spontaneous healing (a 'regression to the mean' of the health state) of the patient. We already know about the confirmation bias; the placebo effect is the substantial positive influence that the patient's psychological expectation has on the healing process; and the timespan of illnesses (with the sole exception of chronic and terminal diseases) ensures that patients spontaneously heal over time.

Because of those three factors, 'therapy experience' – the personal experience that a therapist has with a certain treatment – is a very poor indicator of the effectiveness of that treatment. Firstly, the therapist is influenced by the confirmation bias and is therefore unconsciously more receptive to 'evidence' that supports the effectiveness of the therapy, than evidence that undermines its effectiveness. Secondly, the patient believes in the effectiveness of the therapy and therefore benefits from a strong placebo effect. Finally, the passing of time (spontaneous healing) does the rest. This is why many people strongly believe in the effectiveness of homeopathy, acupuncture, and even ear candling therapy, and that the therapist, often in good conscience, believes that the therapy is effective based on 'successful' experiences with the therapy.

Randomized double-blind trials with control group

To avoid these pitfalls, therapies and medication should be tested in 'randomized double-blind trials with control group'. A control group is a group of test subjects that is given a placebo to compare with a group receiving the actual treatment and check whether the treatment or medication is effective over and above the placebo effect. The allocation of patients to the treatment group and the control group, is done randomly so as to avoid a biased selection of patients (for example, by putting the hopeful cases in the treatment group).

Finally, the research is double-blind. This means that not only do the patients not know whether they receive a placebo or the real medication or treatment (obviously), but neither does the researcher who interprets the results. The researchers should not know which patients are in which group to avoid being biased by their confirmation bias – which would typically lead them to look for confirming evidence that the therapy works by interpreting the results from the therapy group in a more positive light than the results of the control or placebo group. When popular 'alternative' therapies are tested in such a way, they often show no effect whatsoever over and above the placebo effect.

What about traditional medicine?

Whilst the argument may seem reasonable that certain forms of therapy are part of age-old traditions (think of all sorts of alternative Eastern therapies) and must therefore be effective, it is actually not a good argument for the therapy's effectiveness. Just because something has been carried out for centuries, does not mean that it is beneficial or even that it is not harmful. Think of bloodletting, for example. This form of medical treatment was applied in the West for more than two millennia, from Greek Antiquity up until the 19th century. People throughout the centuries thought that all kinds of diseases could be cured by draining blood (and healers also found strong indications for this, misled as they were by their own therapy experience, distorted by the confirmation bias, the placebo effect, and spontaneous healing).

Draining (sometimes a few liters!) of blood, we now know, is not only ineffective in curing diseases, it is often harmful. Bleeding a weakened body is not exactly the best strategy to cure someone who is ill. George Washington, the first president of the United States, likely died due to bloodletting (and not the disease for which he was being treated, namely laryngitis). Remember how we defined critical thinking in chapter 2 as rational and autonomous thinking. Relying on authority or tradition does not lead to truth.

RELIGION

This brings us to a next domain where irrationality thrives. The domain of irrationality par excellence, not only because of the blatant irrationality of some of the beliefs, but also because of the scale on which these beliefs spread. That domain is religion. Religion is a strange phenomenon. It is universal (as far as we know, in all human societies throughout history people have entertained supernatural beliefs) but often comes at an evolutionary cost (with regard to survival and reproduction). These costs range from economic costs such as offering sacrifices to the gods, over reproductive costs (e.g. imposed celibacy), to health costs – sometimes life threatening – when engaging in extreme rituals. Why, then, are these beliefs and practices so common?

The ingredients for supernatural beliefs

Research in the 'cognitive sciences of religion' reveals a series of psychological factors and biases that play an important role in the emergence and transmission of religious ideas. A first important bias is the so-called 'hyperactive agency detection' (Barrett, 2000), which we discussed in chapter 3. We tend to detect the actions of an agent too quickly when interpreting events. This bias led people in all prescientific cultures to view thunder, lightning, solar eclipses, and other natural phenomena as voluntary actions of one or more supernatural beings (gods, spirits, etc.). Our tendency to attribute causes to events, and to make up causes if we cannot uncover the true cause, also plays an important role here.

Moreover, we are intuitive 'dualists'. We tend to perceive mental phenomena (consciousness) as strictly separate from physical or material phenomena. We do so because we have specific and very different innate intuitions about the material world ('folk physics') and the behavior of others ('folk psychology'). These two types of intuitions enable us to navigate our physical or natural environment and our social environment

We use very different principles in both contexts. Our 'folk psychology' is based on our ability to empathize with others. We attribute intentions, thoughts and emotions to others to understand (and predict) their behavior. Of course, we do not do this with regards to the 'behavior' of physical objects. In this context, there are very different principles at play. As a result, we intuitively see the mental as radically different and separate from the material (and we still tend to do so even though we now have neuroscience mapping out the relationship between physical processes of the brain and mental activity). From such a strong dualism, it is not too big of a leap to disconnect the mental from the physical by talking about an immortal soul in a mortal body, and to hypothesize the existence of purely spiritual (immaterial) entities such as ghosts and gods.

Finally, other biases also play a role in forming religious ideas. For example, we seem to have an intuitive preference for 'teleo-functional' explanations (explanations in terms of purposes). When children and non-scientifically educated adults are asked why a certain rock is pointy, they tend to prefer answers such as 'because animals would not sit on them' over answers such as 'by coincidence' or 'because the wind or the rain have shaped the rock' (). This explains why creation stories are intuitively compelling; the world is experienced as the product of a divine architect in which everything is as it is for a reason.

Once these religious beliefs emerge, they spread like a wildfire. Another important reason why we easily take on such irrational beliefs is our ingroup - outgroup bias. We are taught these beliefs by people close to us (at a young age), and we usually take them on without critically questioning them. The emotional ties we develop with these 'sacred' beliefs make us even less likely to scrutinize them.

THE MYTH OF 'HOMO ECONOMICUS'

Irrationality, however, is not confined to these higher spheres. In our everyday life, dealing with practical problems, our biases also often lead us astray. In this context, behavioral economists have overthrown the traditional view in classical economics that economic actors are always rational. Such a rational economic actor is often referred to as a 'Homo economicus', who has well defined preferences and who maximizes their 'utility' (the value or pleasure they get from their investment and consumption choices) in a perfectly rational way.

The reality of how investors and consumers act, however, is far-removed from this rational ideal. What we are willing to spend on a particular product, for example, is usually the result of 'anchoring' (see problem 10 of chapter 2 and the Appendix) and not of a rational consideration regarding how much utility the product will give us (taking into consideration how much utility any other possible purchase for a similar price would give us). Such a consideration, of course, is much too complicated.

In reality, we are typically prepared to pay reasonable prices for the products we want. To determine the reasonableness of these prices, we look at the prevailing prices on the market. When new products are first introduced, there are no price anchors yet. As a result, people feel lost and are often hesitant to buy the product, even when they really want it. For that reason, marketeers devised a clever way to overcome such reticence in the absence of anchors: they introduce several similar products at once, such as a deluxe version, a 'basic' version, and something in between. Doing so, they give consumers an anchor, and the latter predictably purchase the product in between. This happened, for instance, when homemade bread machines were brought to the market.

Another important phenomenon is 'framing' (see problem 11 – chapter 2). A nice example of how consumer behavior can be influenced by framing is described by Dan Ariely (2008) in his book 'Predictably irrational'. The magazine 'The Economist' came up with the following price proposition: Option 1 offered a 1 year subscription to the online version for 65$, Option 2 offered a 1 year print version for 125$, and Option 3 offered a 1 year online and print for ... 125$! Why add an absurd option like option 2? Who would choose the printed version only (option 2) above print and online (option 3) for the same price? Why do they not just give the customer two options? In practice, however, it did not turn out to be so absurd of an idea. When people were given the 3 options, Ariely discovered, 84% opted for option 3 and 16% opted for option 1. If they only received options 1 and 3, 32% opted for option 3 (print and online) and 68% opted for option 1 (online only). By placing options 2 and 3 next to each other, option 3 suddenly seems much more attractive (I will get the online version for free!') So, adding that second option, which no one ever chooses, changes the preferences of the potential customers considerably, and generates more income for The Economist.

There is a whole list of irrational consumer and investor behavior documented by behavioral economists. For instance, there is the Allais paradox (problem 12 – chapter 2), 'loss aversion' (Appendix) and the 'endowment effect', i.e. the fact that we attribute more value to something simply because we own it (Appendix). Our behavior appears to be very far removed from the rational behavior that is assumed in traditional economics. This has important consequences, much like our irrationality in other domains also has important consequences. We will discuss these consequences in chapter 6. In the next chapter, you will learn how you can protect your thinking against irrationality.

SUMMARY

Which biases lead to:

Superstition?
Hyperactive pattern detection, confirmation bias

Conspiracy theories?
Causal reasoning errors, confirmation bias, ingroup-outgroup bias, lack of self-criticism

Pseudoscience?
Lack of self-criticism, confirmation bias

Religion?
Hyperactive agency detection, intuitive dualism, preference for teleo-functional explanations, ingroup-outgroup bias

How do pseudosciences protect their theories against falsification?

- Moving targets
- Built in vagueness

Why is 'therapy experience' unreliable?

- The confirmation bias of the healer
- The placebo effect on the patient
- Spontaneous healing of the patient

What is the scientific protocol for testing a medical treatment or medication?

- Randomized double-blind trials with control group

FURTHER READING

Popper, K. (1963). *Conjectures and refutations: The growth of scientific knowledge.* Routledge & K. Paul Ariely, D. (2008). *Predictably irrational: The hidden forces that shape our decisions.* Harper Perennial Goldacre, B. (2008). *Bad science.* Fourth Estate

Mastering Critical Thinking

How we can guard our thinking against reasoning errors

THREE SOURCES OF REASONING ERRORS

In order to guard our thinking against reasoning errors, we must first be aware of the sources of these biases. In chapter 3 we saw how natural selection equipped us with different thinking systems (system 1 and system 2) and how both systems systematically give rise to reasoning errors. Moreover, we saw that emotions can distort our reasoning. In this chapter, you will learn how you can protect your thinking from irrationality which arises from your intuitive thinking (system 1), from interference with your emotions, and from the confirmation bias and the overconfidence bias (the infamous biases of system 2, our conscious reasoning).

INTUITIVE REASONING ERRORS

There is no off-switch for system 1

How can we guard our thinking against intuitive reasoning errors? First, we must realize that system 1 cannot be switched off. No matter how critical we are, our automatic, intuitive thinking continues to flood our minds with a constant stream of output. Our only protection against intuitive thinking errors is to check our intuitive output with our conscious and reflective thinking (system 2). This requires an effort. We are inclined to sit back and let system 1 do all the work. Remember that Kahneman (2011) calls system 2 the 'lazy controller'. Also, remember Sperber and Mercier (2017) 'argumentative theory of reasoning': our reasoning abilities (system 2 thinking) evolved in order to be able to convince others, not to check up on our own intuitive thinking. Indeed, we are not inclined to reflect upon the reliability of our intuitions.

Even in contexts where people have been trained not to think intuitively about a certain topic, such as in the sciences, it appears that system 1 still frequently distorts human thinking. A good example of this occurred in the research on human evolutionary history. Like most other animal species, we have an (unconscious and intuitive) psychological mechanism that makes a strong distinction between members of our own species and other species. This evolutionary old mechanism has evolved for obvious reasons such as procreation, cooperation, and competition with conspecifics. This mechanism is still present in humans (notably, human languages typically group all other species under the single heading 'animal'). This mechanism prompts us to regard human characteristics as fundamentally different from non-human characteristics.

Even in the context of research on the evolutionary history of humans, where this intuition has been violated by discovering that we share a common ancestor (who lived about 6 million years ago) with chimpanzees and bonobos, the intuition that humans are radically different than other species still appears to have distorted the thinking of paleoanthropologists. Human evolution was initially presented as 'unilineal' and is still perceived by many lay people as such. A unilineal view of human evolution entails that there is a single line of hominid species between the common ancestor that we have with other apes and Homo sapiens (australopithecus - Homo habilis - Homo erectus - Homo sapiens). This view stands in stark contrast to the branching pattern perceived in the evolutionary history of other animal species, in which a given branch typically branches out and some branches become extinct, whilst others, in turn, branch out again.

The reason invoked to explain this difference is that the hominids were able to inhabit very different ecological niches by possessing (primitive forms of) culture. As a result, paleoanthropologists argued, hominids did not split up into different branches – each with specific adaptations to a particular environment – but evolved, as a whole, into what we are now. The underlying assumption (or 'intuition') here is that humans are radically different from other animal species, and that this difference had an impact on our (recent) evolutionary history (De Cruz & De Smedt, 2007).

This unilineal vision of human evolution has, however, been debunked. A rather large variety of hominid species, it turns out, inhabited the world at roughly the same time. All these other hominid species, however, have gone extinct, and Homo sapiens remains the only hominid species alive today. Recently, the remains of a dwarfish upright hominid with a brain size not much larger than that of a chimpanzee were found on the Indonesian island of Flores. They are estimated to have lived about 18,000 years ago, simultaneously with Homo sapiens, and were most likely driven to extinction by our ancestors.

We should, therefore, always be on our guard for the distorting influence of system 1 on our thinking. Even in contexts where we mainly rely on our conscious and reflective thinking processes (system 2), such as in the sciences, system 1 remains active behind the scenes. As I already mentioned before, our intuitive thinking cannot be switched off. All we can do is systematically check our thinking for reasoning errors that were automatically and unconsciously generated by system 1.

Can we never trust our intuition?

This central point of critical thinking (that we should check the output of our intuitive thinking) is at odds with the popular notion that we should 'follow our intuition.' We are advised to follow our gut feeling or inner voice and are often promised that this trust will guide us towards the right decision. In other words, we should have system 1, including the affect heuristics where we make decisions based on emotional reactions rather than based on a well-thought-out cost-benefit analysis, run the show. After what we have learned in the previous chapters, I hope that you understand that this is not an optimal decision-making strategy. Does this mean that we can never rely on our intuition? No! The correct answer is: it depends.

Two kinds of intuitions

'Intuition' refers to two very different sources of beliefs. The first source consists of genetically anchored or innate thought processes. The second consists of automatic but acquired thought processes. With regards to the first source, evolution has equipped us, as previously explained in chapter 3, with fast and frugal reasoning abilities shaped to navigate our environment. Our ancestors did not have the time to think at length about problems they encountered (remember the hominids pondering whether there is still a tiger in the cave – chapter 3). Nor did they have the luxury of possessing even more complex cognitive abilities, because these abilities come with a price tag (there is a trade-off between the accuracy and cost of cognition). Therefore, as we saw in chapter 3, our intuitive thinking is fallible. Moreover, it can lead to reasoning errors because of 'error management' (the fire alarm principle of making more mistakes to avoid costly mistakes – described in chapter 3) and because of a mismatch between the problems for which these intuitions have evolved, and the problems that we are encountering in a modern environment.

Ecological rationality

This does not mean, however, that our intuitive thinking is always misleading. As I pointed out in chapter 3, truth may not be an end in itself for natural selection, but it is usually the best way to ensure the survival and reproduction of an organism, at least when it comes to navigating the natural environment. Recently, there has been a reaction against Kahneman and his colleague Tversky (Tversky & Kahneman, 1974) as well as against other cognitive psychologists who were mainly focused on showing that our intuitive heuristics (the automatic thinking rules of system 1) lead to irrationality. According to the German psychologist Gigerenzer (2000), these heuristics are not 'misleading because they are simple', but rather well-adjusted tools evolved to deal successfully with important 'ecologically relevant' problems.

Heuristics are not the source of irrationality, Gigerenzer argues, but of 'ecological rationality'. They enable us to solve ecologically relevant problems quickly and accurately. Take the 'availability heuristic,' for example. This heuristic, as Kahneman and Tversky discovered, leads to reasoning errors by making us assume, for instance, that deaths from shark attacks occur more frequently than deaths caused by dislodged aircraft parts (chapter 2). Gigerenzer and colleagues, however, rightly point out that the 'availability heuristic' usually produces accurate beliefs because events that are easier to imagine or recall are typically also more common.

Heuristics, Gigerenzer claims, produce true beliefs, at least when they are applied in 'real world' contexts. The reason that Kahneman and colleagues found that our intuitive thinking leads to irrationality, is because they tested our intuitions in an artificial-experimental context designed to make us err! In everyday life, however, our heuristics are generally reliable. Incidentally, Gigerenzer and colleagues refer to Kahenman and colleagues as 'the people are stupid school of thought'.

Does this mean that we can trust system 1 blindly? Of course not. These heuristics, as Gigenerzer knows, are only reliable insofar as they are applied in an ecologically valid context. Hence the term 'ecological rationality', which refers to the ancestral context in which most of human evolution took place. The danger of a mismatch between these heuristics and the context in which they are now put to work, however, only increases as we move away from our ancestral environment.

Think of assessing financial risks, developing theories in quantum physics, or conducting statistical analysis. Our intuitions are no good in these contexts. However, we can usually rely on them in an everyday context. For example, in finding out what is most prevalent in our environment (using the availability heuristic), or who can be trusted in our social environment. Evolutionary psychologists (Cosmides & Tooby, 1992), in this context, argue that we are endowed with a well-adjusted 'cheater detection module'.

So, we should start by becoming aware of the fact that our intuitions automatically generate beliefs. When we notice that we have formed a belief intuitively, we should check whether or not our intuitions are reliable in the context in which they are applied. Are we dealing with a context that is 'ecologically valid'? Or, put differently, is the context in which we apply our intuitions not fundamentally different from our ancestral context? If that is the case, we can usually trust our intuitive thinking. If we apply them in a context that is far removed from this ecological context (e.g. modern sciences, financial markets and statistics), we should turn on system 2 (consciously think it through) and refrain from going with the first thought that comes to mind (our intuitive output).

We should also be aware of the cognitive pitfalls inherent to our intuitive thinking, mainly due to 'error management. When we see patterns and make causal connections, for instance, an alarm bell should go off. It may well be that there is a correlation, perhaps even a causal correlation, but – as we saw in chapter 4 – we also know that we are prone to make such connections too quickly. It is therefore advisable to take a step back, to engage our reflective thinking and to make sure that there is indeed a pattern or causal correlation. Do not be fooled like the instructor at the Israeli aviation base (mentioned in chapter 4)!

The same applies, of course, to those other domains in which we are often irrational (chapter 4). By being aware of these domains we can develop the reflex to critically check our intuitions in these contexts. In short, system 1 cannot be switched off (and fortunately so, life would be unlivable if every action, belief and decision were the product of laborious, conscious and slow cognitive processes) and can in most cases be trusted. In some contexts, however, it makes us predictably irrational. We need to be aware of this and switch to system 2 in these contexts to avoid the cognitive pitfalls inherent to system 1.

Acquired intuitions

Intuitive thinking however, also comes from a second source. Automatic and unconscious thinking processes are not only shaped by natural selection (and genetically anchored), they are also shaped by our experiences. Take the cognitive processes involved when driving a car, for example. Initially, when you first learn to drive, everything happens consciously (and slowly): turning the key, pressing the clutch, switching into first gear, looking in the mirror, etc. Here, system 2 is clearly running the show. After enough practice, however, these actions occur automatically and unconsciously. In other words, system 1 has taken over. Something similar happens when people develop expertise. By repeatedly performing certain cognitive tasks, we can often learn to perform these tasks automatically, unconsciously and accurately.

In his book 'Blink', in which he praises the power of intuitive thinking, Malcolm Gladwell (2005) discusses two salient examples of this process. The first is 'chick sexing': determining the sex of chicks. Distinguishing male from female chicks, it turns out, is difficult when the chicks are only a few days old. Because of the economic incentive to separate male from female chicks as soon as possible, courses exist that teach people to do this quickly and accurately. There is no single characteristic that allows one to determine the chick's sex with complete certainty. Instead, there are various characteristics more often found in chickens than in roosters, and vice versa. Professional 'chick sexers' have so much experience with this task, that they can determine the sex accurately with a single glance. Interestingly, they do this intuitively. They do not consciously make up their mind, which explains of course why they can do the job in a matter of seconds. With years of practice and experience under their belt, these 'chick sexers' have learned to distinguish male from female chicks intuitively.

In his second example, Gladwell (2005) recounts the story of an allegedly antique artwork which was confirmed to be authentic through a series of tests. When an expert saw it, however, he immediately knew that the artwork was a counterfeit. This expert, too, came to this decision not through conscious reasoning, but rather felt this intuitively. Yet again, extensive experience enabled the expert to come to intuitively reliable judgments, since the artwork was later confirmed to have been forged. Intuition from this source is therefore (generally) reliable, at least when these intuitive thought processes are the result of a reliable learning process.

A manual for intuitions

Circling back to the question if we can or cannot trust our intuition. we should proceed as follows. First, we must check the origin of our intuition: Is it an acquired intuition as a result of a learning process, or an innate intuition? If it is an innate intuition, we must ask ourselves whether the context in which we apply the intuition is a context in which our intuitions are generally reliable. Do we apply it in an 'ecologically valid' domain, or a domain to which these intuitions are not attuned? In the latter case, an alarm bell should go off, warning us that following our intuition is not advisable.

EMOTIONS

So much for the 'manual' for dealing with intuitions. Irrationality, however, does not only come from automatic and unconscious thinking processes. Another infamous source of irrationality is our emotions. Our thinking, as discussed in chapter 3, is not (always) isolated from our feelings. Emotions do not only play an important role in the selection of beliefs we take on board through our ingroup-outgroup bias, but also in the selection of beliefs that we refuse to throw overboard. We often develop emotional ties with our beliefs. This is evidently the case with religious beliefs but also, for example, with our political or moral points of view and more generally with opinions that we have publicly defended in the past.

Irrational forms of 'cognitive dissonance reduction'

When we are presented with strong counterevidence to our beliefs (strong enough to get through the filter of the confirmation bias) a state of 'cognitive dissonance' occurs. Our beliefs are not consistent with the information that comes from reality. We usually find this unpleasant. We prefer to see ourselves as rational beings (beings that represent the world accurately). As critical thinkers we should of course eliminate the dissonance by discarding our beliefs or at least adjusting them, but as emotional beings we often refuse to do so. Instead, we engage in a different kind of dissonance reduction. An irrational kind. We do not adapt our beliefs to the outside world, but our perception of the outside world to our beliefs. In other words, because we cherish these beliefs so much, we keep them on board and reduce the dissonance by adapting the interpretation of the facts.

A striking example of this comes (again) from the wonderous world of the sects. A sectarian group in the U.S. believed that there would be a flood that would destroy the whole world in the morning of December 21st, 1954. The members of the sect believed that they would be saved right before the flood by aliens coming to their rescue in a flying saucer from the planet Clarion. Leon Festinger (1957), a psychologist, went over to see what would happen when beliefs in which people are strongly emotionally invested were unmistakably refuted. He observed the sect members standing atop of a hill in California on that faithful morning, ready to board the flying saucer. As you can imagine, nothing happened. No flying saucer and no flood. What would they do?

Some left the sect disillusioned. Most, however, stuck to their beliefs and came up with a special explanation for the facts. According to them, God had decided to save the world at the very last minute, because the small group of believers had 'spread so much light'. A great example of irrational dissonance reduction: the belief is not adapted to the facts, but the facts are interpreted in such a way so that the belief remains intact. And in a much more subtle way (thank goodness!), we are all susceptible to this kind of dissonance reduction.

Think for example of someone who wants to live in an environmentally conscious way, but does not want to sell their polluting car and rationalizes this decision (e.g. if I sell it somebody else will just drive it, or that the greenhouse effect is mainly due to cattle, or that one car does not make a difference, etc.). The same goes for the smoker who wants to live healthily and minimizes the health risks of smoking or the athlete who takes doping but does not want to see himself as a cheater and tells himself that everyone does it.

The psychological mechanism of irrational dissonance reduction can have far-reaching negative consequences for society. Take, for example, the politician who has minimized or denied climate change throughout their life. They will be inclined to be very skeptical of new information about the impact of greenhouse gases on the climate. The same applies to other socio-economic issues where politicians have taken a stand. They tend to dig their heels in when presented with counterevidence.

HOW CAN WE CURB THE CONFIRMATION BIAS?

Our intuitions and emotions are not the only ones to blame for our irrationality, our conscious and reflective thinking also leads us astray. The biases arising from our conscious thinking (system 2) are the confirmation bias and the overconfidence bias following from it. The success of modern science, as we saw in chapter 4 (and will further discuss in chapter 7), can be attributed to the built-in protection mechanisms against this universal human bias. In a similar way, we too can protect our thinking.

We can do so in two ways. Firstly, we can limit the confirmation bias by following the example of Darwin (see chapter 4): by being aware that we are affected by it (we all are!) and by making a conscious effort to look for and record evidence or arguments that would refute our beliefs. In other words, we can play devil's advocate in our own thinking. When we form an opinion or belief, we should not – as we are inclined to do – (only) search for supporting evidence but also for counterevidence.

A second way to curtail our confirmation bias is to surround ourselves with (and listen to) people who think differently. We are not inclined to do that either. Engaging in discussions is not our favorite social activity (we prefer to talk to like-minded people) and yet we all need it to keep our thinking in check. Companies, for example, benefit from a board where disagreements occur and dissenting opinions are expressed.

The wisdom of the crowds

Other people, it turns out, are very good at uncovering the fallacies in our arguments, just as we are very good at exposing the fallacies that have crept into the reasoning of others. Only, we seem to lose most of this ability when it comes to our own beliefs and arguments (see the 'bias blind spot' explained in the Appendix). As a result, groups generally come to more accurate beliefs than individuals. This phenomenon is often referred to as the 'wisdom of the crowds' (Surowiecki, 2004).

At the beginning of the 20th century, Francis Galton, Darwin's nephew, discovered this phenomenon. A large group of people was asked to estimate the weight of an ox. It turned out that the median of their answers was extremely accurate (the median answer had an error margin of less than 1%). The larger the group, the greater the diversity in the group and the more the opinions of the individuals within the group are formed independently of each other, the more accurate the group becomes. The reason for this is that such a form of group thinking corrects for the individual errors and tunnel vision of each of the members of the group.

Large groups consisting of laymen often prove better at making predictions in economic and political contexts than the best experts! It is crucial, however, that the group does not behave as a group. For the wisdom of the crowds to materialize, members should not be allowed to communicate and consequently influence each other. Otherwise, social emotions such as conformism (ingroup bias) take over and the wisdom of the crowds often disappears.

The overconfidence bias

By making our beliefs vulnerable in this way, we also get rid of that other bias which follows from the confirmation bias, namely the overconfidence bias. We are usually much more certain that we are right about something than is justified. Research shows that people who estimate the probability that they are wrong at 1 in 100 are correct only 73% of the time, and even those who are so certain that they estimate the probability of being wrong between 1 in 1000 and 1 a 1.000.000, are only correct 85% of the time (Fischhoff et al., 1977)!

Moreover, people who are generally very confident that they are right, tend to predict much worse than people who are typically less certain. This is not surprising in the light of the confirmation bias. The more certain we feel, the more we succumb to tunnel vision and the more oblivious we become to counterevidence. To remedy this, we must get out of the confines of our own thinking. We must expose our beliefs to the critical gaze of others. Only in this way can we rid ourselves of our unfounded certainties. Only in this way do we tap into the wisdom of the crowds.

The extended mind thesis

That brings me to a more general observation. The great intellectual achievements of Homo sapiens are not so much the product of our 'naked' intellect. We have had our modern sized brains for some 200 000 years. For most of our history as biologically modern human beings, however, this did not get us much further than tending to fire and making rudimentary tools. What made possible the great cultural leap forward, is not so much our brain activity in isolation, but the use of external elements in our thinking. This insight is at the core of the influential 'extended mind thesis' (Clark & Chalmers, 1998). Our minds (or our thought processes) are said to extend beyond the boundaries of our brains. Think, for example. about the way we remember things by writing them down or the way we find our bearings by relying on signposts or our use of calculators to solve complex calculations.

Three levers for our thinking

There are three types of mind-external 'levers' that can scaffold our thinking (bring it to a higher level). The first lever consists of *other minds*. All major scientific discoveries and technological breakthroughs are the product of a collaboration of minds (people). Both a collaboration through time – scientists build on the work of previous generations of scientists – and a collaboration at the same time (scientists work in teams or test their ideas by presenting them to others). The importance of this cooperative form of knowledge acquisition can hardly be overestimated. According to the influential primatologist and psychologist, Michael Tomasello (2009), the most important cognitive ability of humans – and the one that underlies the difference in cultural complexity between humans and other animal species – is our ability to pool knowledge and build on the knowledge of others.

The second lever consists of the so-called '*cognitive artifacts*' (such as logic, mathematics and language) that we have developed. They enable us to look at reality in a completely different way. Mathematics offer us a radically new way to interpret the data we gather from the world. Language does not only enable us to communicate and share our knowledge (supporting the first lever) but also to reflect upon our thinking and therefore be able to question the output of that thinking. To question a representation of the world, one must first be aware that one possesses that representation. Only by having linguistic representations (as opposed to unformulated intuitive representations) can we become aware of those representations and consequently change those.

The third lever, finally, consists of the *instruments* we use to support and enhance our thinking. These tools range from writing with which we radically expand the possibilities of our memory but also support long and complex reasoning processes (think of the use of writing in long calculations), to technological tools with which we extend the reach of our senses (such as telescopes) or perform complex computational operations (such as calculators and computers).

Outsourcing our thinking

So, the true power of our thinking does not reside between our ears but outside of our heads. To think properly, we must involve the outside world. We must appeal to other minds and make use of the cognitive and technological artefacts we have at our disposal. In a certain sense, we must outsource our thinking. A good recent example of this is the success of statistical prediction rules. These are equations in which relevant factors are given a certain statistical weight to arrive at a prediction. For example, an equation was developed to predict the price of wine at an auction, based on the age of the vines and all kinds of climatic factors. Such formulas tend to produce more accurate predictions than the best experts in the world!

Statistical prediction rules enable us to better predict outcomes in a wide range of contexts. For example, to assess the chances of success of a marriage or the chance of recidivism of criminals, but also when making medical diagnoses, assessing credit risk for banks and even to predict the productivity of a job-applicant. With regards to the latter, it turns out that it is better not to invite the applicant for a face-to-face interview because such unstructured job interviews significantly reduce the chance of attracting the best candidate for the job. And in times where these statistical prediction rules are released on ever larger data sets (big data), the accuracy of these predictions only increases.

The 'take away message' is that to think better (read: think more critically), we must acknowledge the limitations of our thinking. We are all susceptible to (the same) cognitive illusions, and we are all inclined to overestimate the probability of being right. A critical thinker is someone who consciously makes an effort to get rid of tunnel vision. Someone who is prepared to scrutinize their beliefs and is always prepared to revise those beliefs in light of new information. This is not something we do spontaneously. It goes against our nature. That is why critical thinking requires a conscious effort. It is a disciplined way of thinking.

Thinking about thinking

The essence of critical thinking is thinking about thinking. We must make a habit of asking ourselves if we can trust our thinking. We need to consider what a belief is based on: intuition or reasoning? Whether there are possible cognitive pitfalls. Whether we are emotionally involved in our beliefs and opinions. Whether we have used all available external levers. In short, we should keep questioning the output of our own thinking. Albert Einstein, one of the greatest thinkers in recent history, reportedly said the following: 'It is not that I am so smart, but I stay with the questions much longer.' That is critical thinking.

SUMMARY

How to protect our thinking against reasoning errors coming from:

Intuitions

Check their origin

- Innate intuitions: only reliable in an ecologically valid context
- Acquired intuitions: usually reliable

Emotions

Be careful not to engage in irrational cognitive dissonance reduction!

Confirmation bias - Overconfidence bias (system 2 biases)

- Be aware of these biases!
- Play devil's advocate in your own thinking
- Surround yourself with people who think differently

What is the 'wisdom of the crowd'?

When a large group of laypeople is asked to estimate something - and when the answers are formulated independently - the median of their answers is typically found to be very close to the correct answer.

What is the 'extended mind thesis'?

Our minds (or our thought processes) extend beyond the confines of our brain.

Which three external levers do we use in our thinking?

1. Other minds
2. Cognitive artifacts
3. Instruments

6 The Importance of Critical Thinking

ARE THERE BENEFICIAL ILLUSIONS?

Now that we have discussed *how* we can think critically, the question remains as to *why* we should think critically. That is not always obvious. Some illusions may be useful. The famous French author, Marcel Proust (1949), thought that we needed to entertain some illusions to make life bearable. In his masterpiece, 'In remembrance of things past', he writes that 'if we are to make reality endurable, we must all nourish a fantasy or two'. Many, I believe, might agree. I have previously mentioned the self-deluding overestimation of one's own abilities of the psychologically healthy person (as opposed to the depressed realist). Marx regarded religion as opium of the people. And what is wrong with a good placebo?

Take homeopathy, for example. If people believe it works and get a nice placebo effect out of it, where is the harm? Should we really expose their illusion? A similar argument can be made for religion. Belief in an afterlife, an immortal soul, and a loving God who watches over us, can be a huge psychological support for people. Who are we to take this from them? Everyone is of course free to believe what they want. Imposing beliefs on people would be the opposite of critical thinking. That would be dogmatic thinking.

Nevertheless, we must be on our guard for illusions. For two reasons. Firstly, they often come with negative consequences. Take alternative medicine, for example. If people turn to it to 'cure' minor diseases that are not life threatening, there is no real problem. However, practitioners of alternative treatments often boast to treat more harmful afflictions. There are currently homeopathic medicines on the market that falsely claim to protect against malaria. Some patients with serious conditions (such as cancer) also prefer alternative therapies to mainstream medicine. Moreover, an alarming number of alternative therapists incite people not to vaccinate themselves and their children. This is based on unfounded rumors that vaccinations are detrimental to health and can lead to autism for example. The anti-vaccination movement does not only put the unvaccinated children at a serious risk, but society at large since it opens the door for the spread of deadly diseases. The price we pay for these illusions comes in the form of human lives. And the dark side of other illusions, such as religion, should also be clear. We will discuss that later.

The second reason why we should be careful with illusions is because illusions - as we have seen in chapter 2 – tend to branch out in our thinking. We want to maintain a coherent worldview, so illusions usually produce more illusions. People who believe in the predictive power of astrology, for example, also appear to be more susceptible to other illusions such as the existence of mediums, psychics, and the potency of 'energy healing'. Even if an illusion would only have positive effects, it is not inconceivable that it makes us more susceptible to illusions that could be harmful. We cannot take nonsense on board selectively. Once we open the door to it, even a little bit, our thinking can rapidly be flooded (Boudry, 2016).

THE IMPACT OF IRRATIONALITY ON THE WORLD

Overconfidence and war

Much more important than the impact of irrationality on our own lives, is the impact of irrationality on the world. Viewed through the lens of critical thinking, many of the major problems in the world appear in a new light. Warfare, for example, is fueled by a strong overconfidence bias on at least one side. If you do not believe that you can win, you generally do not go to war.

According to the historian Geoffrey Blainey (1988), blind optimism is a central component in the build-up to any war. At the start of the first World War, for examples, both camps believed that they would finish the job in a few months and would celebrate Christmas 1914 victoriously at home. In the run up to wars, there is generally a feeling of euphoria and fighting spirit - especially amongst the young men who will go to war. A euphoric feeling that quickly dissipates once the sad spectacle has begun and drags on much longer than expected.

According to the political scientist Dominic Johnson (2004), overconfidence is the cause of many wars. He points out that there is a strong correlation between the form of political decision-making of a country and the chance that it will undertake military actions. In societies in which a political debate precedes decision-making (as is generally the case in democratic societies) and the overconfidence bias of individuals is therefore often tempered, considerably less military action is undertaken than in societies in which this is not the case.

The way in which the protagonists deal with intelligence also plays a role. When the protagonists keep their head cool and let reason prevail, war can often be avoided. Think of the Cold War in which a nuclear Armageddon was avoided by the thoughtful action and communication of Kennedy and Khrushchev and the diplomats from both sides. In both the Vietnam War and the last invasion of Iraq, the Americans were less cautious and fell prey to the overconfidence bias. With respect to the invasion of Iraq perhaps not so much about their military superiority, but about the aftermath of their military actions (the prospects of stabilizing the region).

I believe we are witnessing something similar with Russia's invasion of Ukraine. At the time of finishing this book, the war in Ukraine has just started, so conclusions are somewhat premature. But Putin, the Russian president and autocrat, seems to have underestimated Ukraine's military resistance and seems overconfident in his assessment of how that military operation will play out for Russia and his regime.

The ingredients of financial crises

The same 'positive illusions' – overconfidence and the illusion of control – also impact other areas, such as the financial world, for instance.

Thinking that you can predict the market turns out to be an illusion that so-called stock exchange experts are very susceptible to. Despite complex models and strategies, experts seem to have no clue what the market will do. The economist Burton Malkiel puts it rather forcefully. According to Malkiel (2003), 'a monkey throwing darts at the financial pages of the newspaper can put together an equally good portfolio as financial experts'!

Yet we continue to 'detect' patterns in the movements of the market (remember our hyperactive pattern detection). We are often also very impressed by the big winners of the stock market, those who beat the market by a large margin (sometimes several years in a row) and take their success as evidence that the evolution of the market can indeed be predicted. But for every winner there are many losers and we grossly underestimate the role that luck plays in this. Large statistical studies reveal that the correlation between the performances of top traders over successive years is as good as zero!

Nevertheless, the false belief of control over the market, produced by our overconfidence bias, hyperactive pattern detection and the success bias (the fact that we hear more about successful than about unsuccessful investors and are inclined to underestimate the factor of luck), makes investors blind to the risks. This, in turn, creates market bubbles and crises. Notably, a UN report on the most recent financial crisis identified the illusion of risk-free profits as its main cause.

It is also problematic, of course, that the 'incentives' created by these financial institutions (the bonus system) promote short-term profit-making and risk-taking. Human nature (and especially male human nature, women turn out to be wiser investors) does the rest. Sadly enough, the price for the irrationality of investors and financial institutions befalls on society as a whole.

WHAT ABOUT RELIGION?

So, are illusions never beneficial? Traditionally, one domain of illusions was considered beneficial and even necessary. That domain is religion. Napoleon, who was not religiously inclined, nevertheless thought that religion was absolutely necessary for the maintenance of social order. So did Georges Washington, the first president of the United States. He reportedly claimed that: 'religion and morality are the essential pillars of civil society'. Religion was traditionally regarded as the foundation of morality, and many people still believe this to be the case. In the American 'bible belt' (a deeply religious region in the South and Mid-West of the U.S.) atheists are the least trusted group of people. They are less trusted than any other minority in the US, including Muslims, and this after 9/11. Since atheists do not believe in God, many religious people think that they have no reason to act morally.

Religion and morality

In chapter 4 we looked at the cognitive underpinnings of religion. We talked about 'hyperactive agency detection', intuitive dualism and a preference for teleo-functional explanations. The reason why every human society possessed religious beliefs throughout history has nothing to do with morality. And for the vast majority of human history, religious beliefs were not linked to moral rules.

The animistic religions in hunter-gatherer societies typically do not impose moral norms. Similarly, in the Greek and Roman polytheistic religions, the gods were neither moral examples nor were they considered to oversee whether the faithful behaved morally. Morality, it turns out, was integrated in religions somewhat recently. According to the psychologist Ara Norenzayan (2013), moral norms were introduced in religions because groups with such moral religions were better able to maintain internal harmony and cooperation. He argues that the integration of morality in religions made harmonious cooperation possible in ever larger groups. Punishing gods, Norenzayan thinks, played a crucial role in keeping together groups that grew ever larger. The rise of large-scale societies went hand in hand with the rise of what Norenzayan calls 'big moral gods' (the powerful, moralizing and punishing gods in religions such as Judaism, Christianity, and Islam).

But the moral concern of those 'big gods' comes with a dark side. Religious moral norms often work in two directions. On the one hand they increase cooperation, harmony, and altruism within the group. But on the other hand, they often increase hostility towards other groups. This is not a coincidence. The cultural success of these 'moral' religions is as much the product of strengthening the ties within the group as it is the product of enhancing the competition with other groups.

Think of the two major monotheistic religions of our time: Christianity and Islam. The history of these religions is filled with religious wars, conquests and proselytism (the 'conversion' of non-believers). Their considerable cultural success – more than half of the world's population is either Christian or Muslim – is not only due to the harmony they create within the group, but also (perhaps even more so) due to the intolerance they harbor with regards to other groups with other religious beliefs.

So, when we talk about religion and morality, we must always keep in mind that the moral norms that religions propagate are mainly – or at least originally – focused on the relationships within the group, and that this often leads to conflict between different groups with different religions (which create harmony within their respective groups in a different way). Moreover, religion often hinders moral progress. By holding on to ancient texts and regulations, the moral norms that religions propagate are not so easily adapted or improved. Consider, for example, the position of the church or of Islamic scholars on homosexuality.

CRITICAL THINKING AND MORAL PROGRESS

True moral progress actually comes from rational, critical thinking. This may sound strange. What does rationality have to do with morality? Does cool, dispassionate reasoning not lead to immoral behavior? Think of Nazism or the ruthless 'Homo economicus' who is driven by greed. Rationality in itself, it is true, does not automatically lead to moral behavior. It is amoral: neither moral nor immoral. Yet it is precisely our reasoning ability and our capacity for critical thinking that drives the process of moral progress.

The influential Australian moral philosopher, Peter Singer (1993), calls this the *'escalator of reason on morality'*. By reasoning about morality we come to moral behavior that is far removed from the type of behavior for which natural selection has equipped us with moral intuitions. Indeed, morality evolved only for cooperation within the group. Reason – in contrast to religion that generally only strengthens this ingroup - outgroup bias – shows us that there is no fundamental reason why moral conduct should be limited to interactions with fellow group members. Therefore, Singer claims, we have been able to 'expand our moral circle'.

In some cases, reason may even overrule certain harmful moral intuitions. A good example of this is the intuitive moral aversion to homosexuality (prevalent in many societies in the past). The first modern Western thinker to oppose this was the British moral philosopher Jeremy Bentham (18th century). He did so by thinking rationally and ignoring his emotional reactions and intuitions. Bentham argued that homosexuality does not cause any harm and that banning homosexuality brings suffering. So, he concluded that it should be allowed, opposing the Christian tradition that strongly condemned homosexuality because it was 'unnatural and contrary to God's will'.

The natural selection of moral intuitions

Remember how our innate cognitive intuitions have evolved for the sole purpose of enabling our distant ancestors to survive and reproduce. The same goes for our moral intuitions. These moral intuitions evolved to do so by strengthening harmony and cooperation in the small hunter-gatherer bands in which our ancestors lived *and* by not extending that courtesy to rivaling bands. Those intuitions do not always yield desirable results, especially in our modern context.

Our ingroup - outgroup bias, which we have already discussed, causes racism and large-scale wars in the modern context where people of different races live together and groups (and coalitions between groups) become increasingly larger. As mentioned above, we easily succumb to unfounded intuitive aversions against certain forms of behavior such as homosexuality. Finally, we are inclined to violently punish the lack of conformism of group members (for example when violating conventional taboos). Human nature is what it is, and it is certainly not perfect. The good news is that we can improve ourselves thanks to our ability to reflect on these practices and think critically.

Four centuries of moral progress

By thinking rationally about morality, we can make moral progress. From the moment that philosophers began to reflect on morality (autonomously and rationally) - after the Middle Ages during which morality was the exclusive domain of religion - we see a huge wave of moral progress that is still ongoing today. Compare Europe in the 16th century with our society today. In the 16th century women were burnt at the stake because they were suspected of witchcraft, religious dissidents were tortured and murdered and slavery was an institutionalized reality. It was also generally accepted that non-white races and women were inferior and that people who adhered to a different (or no) religion deserved an eternal afterlife in hell, where they would be subjected to the most gruesome torture practices.

So, we must get rid of the image of cold reason versus warm emotion and the idea that rational thinking cannot be reconciled with empathic and moral action. It is our rational and critical thinking that has greatly increased the scope of our empathy and has made this world a much better place. And it is still with rational reflection that philosophers question the status quo today and strive for moral progress.

Think, for example, of animal rights. Philosophers - such as Singer (see above) – refer to research about the emotional and cognitive faculties of animals to argue that animals should also be included in our 'moral circle'. These philosophers do not (typically) argue for animal rights based on emotional considerations. Such considerations would not get us very far. The cuddly panda bear and the elegant dolphin might get some sympathy, but what with less attractive species? Of course, they should not pay the price for the fact that they do not meet our aesthetic standards.

CRITICAL THINKING AND PROGRESS IN GENERAL

The importance of critical thinking, however, goes beyond the domain of morality. Throughout history there has been a struggle between critical and dogmatic thinking. In the history of Western thought, there have been two major breakthroughs of critical thinking. The first came with the birth of philosophy in ancient Greece, the second with the advent of modernity in the Renaissance. In both cases dogmatic thinking was replaced by critical thinking and in both cases the consequences would be far-reaching.

In ancient Greece, for the first time in history, people attempted to understand the world by thinking autonomously and rationally, refusing to rely any longer on mythological stories. Philosophy was born: a way of thinking that questioned everything and formulated answers by advancing rational arguments. Socrates, who is considered to be the founding father of Western philosophy, said of himself that he was the wisest man of Athens, since he was the only one who knew one thing, namely that he knew nothing. For the first time in history, everything that was passed on by tradition was questioned. Because of this critical attitude, society would undergo radical changes. The standard of living was raised because there were more technological innovations and philosophers started thinking about how to organize a just society. As a result, Athens in the 5th and 4th century BC was already experimenting with (admittedly not fully inclusive) forms of democracy.

Something similar happened roughly 2000 years later when the Western world awoke from a millennium of dogmatic thinking dominated by Christianity (the Middle-Ages). Authority and tradition were questioned once again, and critical thinking could revive. Here too, the consequences were far-reaching. From modern philosophy - as the philosophy of the 17th and 18th century is called - the (modern) sciences developed, as well as modern political and moral philosophy. Saying that we owe almost everything to this revival of human thought, is no exaggeration. In addition to all the technological innovations and the exponentially increased standard of living, we owe our freedom and rights to the courageous actions of several great thinkers who put reason above tradition.

Irrationality, or even simply the absence of rational and critical reflection, is not innocent. It is irrationality that causes people to wage war in the name of a God or in the name of some utopian ideology (such as Nazism, communism or militant nationalism). And it is a similar lack of rational thinking about morality that makes people surrender blindly to their ingroup - outgroup bias and turn against other groups simply because they are different.

Bad thinking leads to bad outcomes. A better world follows from better thinking. And it does not take much time. Only two centuries of critical thinking separate the contemporary period (19th – 21st century) from the end of the Middle Ages in the Renaissance. In that period, democracy replaced theocracy, slavery was abolished, women acquired the same rights as men and racist ideologies were largely abandoned. We also live much longer lives (on average) in much better conditions and the world has never been as peaceful as it is today (even though it may not always seem that way). The chance that you die at the hand of another person has never been as small as it is today. Humans never had it as good as they have it today. And that is the merit of several generations of critical thinkers.

THE MAJOR CHALLENGES OF TODAY

Today we face major challenges. For the first time in our history, we have the means to wipe ourselves off the face of the planet (and with us many other animal species). For the first time we must deal with people from foreign groups on the other side of the globe (the global economy). And for the first time we also need to work together on a global scale to ensure a prosperous future for all (climate change).

These are challenges for which natural selection has not equipped us. Our social emotions and intuitions are formidable obstacles in this context. Pessimists think that humanity is headed towards its tragic conclusion, optimists argue that 'necessity is the mother of invention' and that we will successfully meet these challenges. Personally, I side with the optimists. But of one thing we can be certain: whether the pessimists or the optimists will be right depends on one thing and one thing only: will critical thinking prevail over irrationality?

Our thinking is both our greatest asset and our biggest threat. Rational thinking is not merely a matter of intellectual preference or even of self-interest. It is primarily a question of responsibility. A world dominated by irrational thinking is a world of conflict and destruction. A world dominated by rational thinking is a world of harmony and progress. History shows this time and time again.

A LASTING STRUGGLE

It is important to remember that critical thinking is not a spontaneous way of thinking. It is a disciplined way of thinking that we must learn. After the first major emergence of critical thinking in classical antiquity, human thinking reverted into dogmatism during the Middle-Ages. The same can happen today. It is a lasting struggle. We must remain vigilant and protect the achievements of rational thinking (such as the human rights that followed from the modern political and moral philosophy) and continue the positive trend onwards.

Critical thinking, I want to add, is not a uniquely or typically Western or modern way of thinking. And it is certainly not a cold, non-empathic way of thinking. It is a way of thinking that everyone, regardless of their cultural background, can participate in. It was developed in very different cultural contexts. And it is a way of thinking that - as pointed out above - brings progress (both morally and in terms of living standards). We bear the responsibility for the well-being of life on this planet and for future generations because their fate is in our hands. And their fate will be sealed by the quality of our thinking. That is the importance of critical thinking.

SUMMARY

What is the impact of irrationality on the world?

- Overconfidence can lead to war and financial crises.
- Religion can lead to group conflict and is a brake on moral progress.

What leads to moral progress?

Critical, rational thinking

Why?

Singer's escalator effect of reason on morality:

- When we reason about morality we arrive at norms and behavior that are far-removed from the kind of behavior for which our moral intuitions evolved.
- According to Singer, upon reflection on norms and practices we:
  - Expand our moral circle
  - Rid ourselves of unfounded intuitive aversions (like the aversion against homosexuality)

7 The Importance and Reliability of Science

On the scientific method and the demarcation criterion

As I explained in chapter 1, one of the learning goals of this book is to reflect on what characterizes science and what constitutes the essence and the importance of scientific methodology. To do so, we did not (as is typically done in a course of philosophy of science) take a historical perspective and discuss what the most important philosophers of science have to say about science. Our starting point, instead, was human reasoning. We saw how, why and when our spontaneous thinking misleads us and how we can guard ourselves against reasoning errors. In chapter 4, we also saw that the scientific context protects against these reasoning errors and that this underlies the success of science. In this last chapter, I will elaborate on this.

THE SCIENTIFIC METHOD

In the first part of this chapter, we will consider the following question: which aspects of the scientific method make the sciences reliable (or at least more reliable than pseudosciences)? Before we take on this question, we must get a sense of what the scientific method is. It is debatable, however, whether it makes sense to talk about *a single* scientific method at all. We associate the scientific method with conducting empirical, experimental, quantifiable research with the aim of predicting and understanding a domain of reality. But none of these oft-cited features of the scientific method are present in all of the sciences. Formal sciences are not empirical. Evolutionary biology and astrophysics are not experimental. Psychology, sociology and anthropology are often not quantifiable. Biology typically does not make predictions.

HUMAN AND NATURAL SCIENCES

So, we cannot speak of a single scientific method. The domains of the different sciences vary too widely. The difference between sciences is so important that some philosophers and scientists even question whether it makes sense to place all those different attempts at understanding a domain of reality under the same denominator of 'science'. Leaving aside the formal sciences (mathematics and logic), there are two distinct families of (empirical) sciences. Those are the social or human sciences (such as history, psychology, economics, sociology, anthropology, ...) and the natural sciences (such as physics, chemistry, astronomy, geology, biology, ...).

Different objects, different aims

The human sciences focus on human thinking, acting and interacting. The natural sciences focus on the physical and natural world. The objects of both types of sciences are very different. In the natural sciences, scientists study things like quarks, electrons, atoms, molecules, tectonic plates, genes, etc. In the human sciences, they study human thinking and acting (e.g. in psychology) and the interaction between people (e.g. in economics and sociology).

According to the 19th century philosopher Wilhelm Dilthey (1883/1989), this means that both types of sciences also have a very different goal. The natural sciences seek to 'explain' ('Erklären'): to describe the world in terms of cause and effect and their underlying laws (e.g. in the atmosphere of the earth, an object accelerates in vacuum at 9.81 m/s due to the gravity of our planet). The social sciences, on the other hand, aim to 'understand' ('Verstehen'). What led to the French revolution is impossible to explain in terms of external, universal laws, according to Dilthey. To do so you must place yourself in the shoes of the actors of that historical episode (their thinking and feeling). It requires a subjective understanding.

In pointing this out, Dilthey criticized so-called 'positivists' such as Auguste Comte who aimed to provide the social sciences with the same quantitative method(s) as the natural sciences and sought to discover general laws within the domains of the human sciences. The central point that Dilthey makes is that we can explain physical reality with basic entities such as atoms and molecules and their lawful interaction, but not human reality. After all, we cannot explain the French revolution by invoking neuronal activity in the brains of the actors carrying out the event. Therefore, according to Dilthey, the natural and human sciences do not only have a completely different object, but also a very different method and very different aspirations.

For starters, the human sciences cannot accurately predict phenomena (as is the case in the natural sciences). Stars, planets, atoms, molecules and genes behave in a lawful (and therefore predictable) way. Human actors do not. We can predict an eclipse very precisely but cannot do the same for a political revolution or a financial crisis. At most, we can identify a series of factors that increase the chance of these events happening.

Moreover, both kinds of sciences are looking for something else. Natural sciences attempt to expose natural laws, social sciences are looking for generalizations. An object in the Earth's atmosphere will always fall with an acceleration of 9.81 m / s (in vacuum), a group of people living in poverty under a cruel dictator will not always start a revolution (and it is certainly not possible to predict exactly when that revolution will erupt).

Looping effects in the human sciences

Finally, there is another important difference between the natural and human sciences. In the natural sciences there is no interaction between the theory and its object. In the human sciences there often is. The philosopher of science Ian Hacking (1995) calls this the 'looping effect': a theory can influence its object in the human sciences (i.e. the people it describes). The reason for this is that a theory in the human sciences can inform the people it describes and that this can influence their thinking and actions.

A striking example of this occurred in psychology. Up until the nineteenth century, women were often diagnosed with 'hysteria': a mental illness that caused all kinds of symptoms such as anxiety, fainting, insomnia and irritability. The disease, it was thought, stemmed from the uterus (hence "hysteria", derived from 'uterus') and to treat the affliction, physicians often removed the organ. Today we known that the uterus does not cause mental disorders and the disease is no longer recognized. In previous centuries, however, many women who suspected they might be affected by hysterical disorders displayed these symptoms. We see the same thing happening with other mental disorders, such as 'multiple personality disorder'.

Looping effects also occur in the social sciences. Here too, a theory can influence its object (in this case: society). Karl Marx described the contract between industry owner and laborer as a contract that was not concluded between two free parties but within a power relationship in which the worker has no choice and is therefore not a free party. According to Marx, the laborer was forced to work for a subsistence wage to survive and the industry owner took full advantage of this. By describing the economic relationships in the society of his time, however, Marx would change these relationships (and the society he described). Communist revolutions ensued and different societies emerged.

A SELF-CORRECTING PROCESS

The natural and human sciences therefore differ greatly from each other, both in terms of method, purpose and impact on the object of their study. According to some, the term social sciences is an oxymoron, a contradiction in terms – because, they argue, society can never be described scientifically. The natural sciences are commonly referred to as 'the hard sciences' and human sciences as 'the soft sciences'. In Dutch people commonly talk about 'exact sciences' when they refer to natural sciences (as opposed to human sciences).

Yet, (good) human sciences and (good) natural sciences have a very important characteristic in common: namely that they are 'self-correcting'. That is the essence of science, according to the

.amous astrophysicist Carl Sagan (1980). What Sagan means, is that the methodology and context (of both natural and human sciences) correct for the mistakes made by scientists. How does that work?

HOW IS SCIENCE PROTECTED AGAINST THE REASONING ERRORS OF SCIENTISTS?

Just like everyone else, scientists are susceptible to the reasoning errors we have discussed. Remember the paleoanthropologists in chapter 5. Fortunately, the quality of the sciences does not depend so much on the quality of the scientists but on the quality of the methodology and the framework within which science is conducted. They operate in such a way that they protect against the reasoning errors and biases to which the human mind is susceptible.

First, the scientific methods protect against intuitive reasoning errors (system 1 reasoning errors). They make extensive use of cognitive artifacts (see chapter 5). These cognitive artifacts, such as mathematics, logic and statistics, do not only radically extend the scope of sciences (without mathematics, Newtonian physics let alone Einstein's theory of relativity would not have been within our reach), they also protect against intuitive reasoning errors, such as the belief bias, gambler's fallacy, hyperactive pattern detection, chance blindness, base rate fallacy, availability bias, etc. By using mathematical and statistical models and computations, scientists do not succumb to their biased intuitions.

Secondly, the scientific framework and context protect against the pervasive biases of system 2. It protects against the confirmation bias (as well as irrational strategies for cognitive dissonance reduction): through 'peer review', which ensures that each theory is critically screened for errors by other scientists before it is published. The scientific context of open criticism also contributes to this. Scientists do not lack motivation to try to revise the influential theories of their time (or at least refine them). The physicist who refutes Einstein's theory, as I mentioned before, goes in the history books.

On the other hand, the scientific framework protects against the overconfidence bias. Experimental results must be reproducible and often the same experiments are carried out by other researchers to check whether the same results are in fact reached. In doing so, scientists want to ensure that these results are robust and not distorted by statistical anomalies (and an overconfident scientist).

Moreover, scientists engage in so-called 'meta-analyses'. They pool all studies on a given phenomenon to reach more robust conclusions and to check whether the results of certain studies deviate strongly from the other studies (a good sign that they are flawed and should be discarded). In this way, they can weed out distorted results (for example, because certain studies had a sample that was too small and therefore not representative).

Finally, scientists will not only publish the results of their research, but also the precise methodology that they have used to arrive at those results. In this way other scientists can critically analyze the hypotheses that are put forward and they can raise problems with the methodology and the interpretation. In short, the context and procedures of 'the scientific game' ensure that theories are made vulnerable and can be corrected – if needed – by other scientists. This makes science a self-correcting process.

The scientific framework also protects the sciences against emotional distortion. While group thinking or the bandwagon effect undoubtedly affects research groups (just like other groups of people), their distorting effect is limited by the existence of rival research groups. A theory from a research group at a Dutch university may not be subjected to harsh criticism internally but can be critically evaluated by a Chinese research team. They do indeed have access to the data and methodology on which the research is based and there are objective measures for the success of a theory (reproducibility of results, validity of the inferences made, simplicity of interpretation – remember Occam's razor? – coherence with other empirically supported theories, etc.).

And again: scientists do not lack the motivation to shed a critical light on existing theories. It is that rivalry (between researchers and research teams) that underlies the power of the sciences. That is the beauty of critical / rational / scientific thinking: it is the only universal kind of thinking. Different cultures have different values, customs and beliefs, but logic, mathematics and statistics are the same everywhere and everyone can expose reasoning errors and help build better theories.

The power of the community

The more scientists join the ranks and try to improve each other's theories, the faster we move forward. The power of the sciences does not reside in the genius of individual scientists but in the size and diversity of the scientific community and in its self-correcting nature.

The only condition for scientific progress is that new theories, including the methodology, interpretation and the data through which they came about, are shared with the entire scientific community and are thus subjected to possible criticism. We do not tend to subject our pet theories to harsh criticism spontaneously and precisely this is missing in pseudosciences. In the latter, advocates of a theory often stick to it dogmatically and surround themselves with like-minded people.

THE IMPORTANCE OF SCIENTIFIC PROGRESS

The importance of scientific progress can hardly be overestimated. Rational thinking in general and science in particular, is the driving force behind improving living conditions throughout history. Most recently, the explosion of the modern sciences in the 20th century produced an unprecedented improvement in the living conditions and longevity. In the year 1900 the average life expectancy in Western Europe was around 46 years, today it is above 80 years (and globally around 73 years). In the year 1910 more than 74% of the world population lived in extreme poverty and in 1980 that was still the case for more than 43%. Today it is less than 10%.[1]

The sciences do not only have a crucial role in improving our living

[1] Sources: https://ourworldindata.org/life-expectancy; https://ourworldindata.org/extreme-poverty.

standards and conditions, but also in improving society. Herein lies the importance of the social or human sciences. To improve society, we must start by understanding its ingredients: the people that populate societies. In other words, to improve society we need to develop the human sciences. The insights of social psychology, for example, are invaluable for social problems such as multicultural integration, radicalization and populism.

Ironically, the human sciences are still in their infancy compared to the natural sciences. While the latter are of course very valuable and quench our thirst for knowledge about the world, the former are crucial for the future of our species, and by extension of countless other species. Some years ago, at the University of Oxford, there was a symposium on 'existential risk', namely the risk that humans will eradicate themselves. The participating researchers (from all kinds of scientific branches) were asked to estimate the chance that humanity will have destroyed itself by the year 2100. The median of their answers was 19%! (Bostrom, 2013) Personally, I am a lot more optimistic (perhaps it is not surprising that researchers dealing with existential risk are pessimistically inclined), but it shows the enormous importance of reaching a better understanding of what underlies the problems and challenges in our society in order to tackle them better.

SELF-CENSORSHIP IN THE HUMAN SCIENCES

Reaching a better understanding of humans and society, however, is often impeded. An important reason for the relative lag of the social sciences is that there are taboo subjects or issues that scientists often steer clear of in fear of causing negative social consequences. A good illustration of this occurred in the 1970s. Edward Wilson, an American biologist who had until then mainly studied ants, suggested that human social behavior could also be explained by analyzing the evolutionary past of our species and was therefore largely determined by our genes. Wilson was called a racist, sexist and even Nazi sympathizer. At an academic presentation of his work, students dragged Wilson from the stage and poured a jug of water over his head while chanting 'Racist Wilson you can't hide, we charge you with genocide!' [2]

What was it, you might wonder, that did provoke such an extreme reaction? Well, the consensus in the social sciences was that the social environment (and not genetics) determined human behavior and the reasons for this stance were not purely scientific. It was a reaction against the blatant racism and sexism of the 19th century where it was commonly thought that there were important genetic differences between the different races and sexes with regards to intelligence. Something that turned out to be false. Nevertheless, it remained taboo for most of the 20th century to study human behavior and qualities from a genetic, biological or evolutionary perspective. However, understanding human social nature and its evolutionary origins is an important piece of the puzzle in understanding society and meeting the important societal challenges standing in the way of a peaceful, harmonious global society.

THE DEMARCATION CRITERION

The sciences must therefore be inclusive. They must not engage in self-censorship in the search for truth. That does not mean, however, that everything must be admitted. The question is not only about which research questions and hypotheses should be admitted because they provide valuable insights. It is also about which theories should not be admitted because they are completely unfounded (pseudoscientific). Most people agree that we should not consider astrology as a legitimate science and that chemotherapy is more effective in the fight against cancer than energy healing. But the question remains, how strict must we be? And – equally importantly – based on which criteria do we distinguish legitimate from pseudoscience?

Popper's falsifiability

Philosophers of science have addressed this question and proposed

[2] Source: https://www.nytimes.com/2021/12/27/science/eo-wilson-dead.html

demarcation criteria, i.e. criteria to distinguish science from pseudoscience. The most influential demarcation criterion, as we saw in chapter 4, is Karl Popper's (1963) criterion of 'falsifiability'. According to Popper, a theory is only scientific if it is testable. This means that it must in principle be possible to refute the theory based on observation (which of course does not mean that the theory will be refuted!).

With his criterion, Popper went against the traditional demarcation criterion of his time, namely: 'verifiability'. According to the latter, a theory is only scientific when it is shown by observation that the hypothesis is right. According to Popper, this is impossible: no theory has ever been verified. Scientists do not prove anything with absolute certainty. The reason is simple, we can never observe everything and so there is always the possibility that future observations will falsify a theory.

But Popper also went against another criterion: confirmation. It is not because a hypothesis is supported by observation that it is scientific. Whereas verifiability is too strong a criterion, confirmation is too weak: many pseudosciences (such as astrology) are 'supported' by a long list of confirming observations. Only, these theories are rarely tested: astrologists do not attempt to refute their theories (being under the spell of the confirmation bias). In this sense, Popper's criterion protects against the confirmation bias by explicitly inciting scientists not to look for confirmation for their hypotheses but instead to look for counterevidence and counterarguments.

According to Popper (1963), scientific progress is driven by: 'conjectures and refutations'. Every time a theory or an aspect of a theory is refuted, another one takes its place, which in turn is tested. In this way theories improve over time. The consequence, however, is that according to Popper we can never state with absolute certainty that a theory is true. If we do so, we slip into dogmatic thinking, the opposite of scientific thinking. Scientific investigation, according to Popper, is – or at least should be – a constant attempt to refute existing theories, not one of seeking additional evidence for theories.

Popper's demarcation criterion may be particularly influential, but that does not mean that it was not criticized. Fellow scientists and philosophers of science exposed a series of problems with the criterion. First and foremost, it appears that scientists do not practice science in the way Popper envisions it (and in the way that his criterion requires). They do not just throw a theory overboard when confronted with counterproof. They will often come up with ad hoc hypotheses to explain the observed anomaly.

For example, when it turned out that the orbit of Uranus around the sun was inconsistent with Newtonian laws, Newton's theory was not discarded. Scientists assumed that there must be another planet that affected Uranus's orbit and this indeed turned out to be the case. Astronomers peered into the solar systems with improving telescopes and found that planet: Neptune. Falsifying Newtonian physics was not called for and this is often the case. Often a theory with a large explanatory scope should not be discarded when we are confronted with inconsistent observations.

Another point Popper's critics made against his demarcation criterion is that pseudoscientists sometimes make falsifiable claims, such as astrologers who make testable predictions about personality and the future based on horoscopes. This does not make these predictions scientific. Of course, when such pseudosciences engage in risky predictions, they are typically falsified sooner rather than later. As such, this is not so much of a problem for Popper's criterion.

Feyerabend's epistemological anarchism

A more fundamental criticism, however, came from philosopher of science Paul Feyerabend (1970), who considers himself to be an epistemological anarchist. According to Feyerabend there is not one correct way to understand reality, but many different and valuable ways. The world is so much more complex than presented in scientific models and theories, and when we limit ourselves to a single perspective on reality (a scientific worldview), we are left with an impoverished worldview.

According to Feyerabend, we should therefore never limit ourselves to one method, both within the sciences and in general. His principle is: 'anything goes'! He is therefore radically opposed to a demarcation criterion. Such a criterion, he argues, prevents new knowledge from being acquired and thus prevents knowledge from progressing. Major scientific breakthroughs, Feyerabend argues, came about precisely because scientists broke the rules of their time. The Copernican revolution, the atomic model of Bohr, ... they all came about, according to Feyerabend, because scientists ignored the methodological rules of their time. Rules prevent progress, Feyerabend claims. We should let everything bloom instead of constantly weeding out 'bad' theories.

In the forceful words of Feyerabend (1970, p. 11): "It is thus possible to create a tradition that is held together by strict rules and that is successful to some extent. But is it desirable to support such a tradition to the exclusion of everything else? Should we transfer to it the sole rights for dealing in knowledge, so that any result that has been obtained by other methods is at once ruled out of court? This is the question I intend to address in the present essay. And to this question my answer will be a firm and resounding NO."

Postmodern constructivism

Feyerabend's view fits in the context of postmodernism. According to postmodern thinkers, there are no objective facts, only constructions and interpretations. Scientists are therefore not discoverers of reality but rather sculptors of reality. Science is no better or more accurate than magic or voodoo, just a different perspective, a different construction and as such there is no reason to admit the former, while rejecting the latter.

Feyerabend strives for what he calls the separation of state and science. Analogically to the separation of state and religion, i.e. a state in which no religion is imposed on its citizens, no scientific worldview should be imposed on its citizens. We should be free, Feyerabend claims, to choose to give our children an education in voodoo, rain dancing, astrology, and/or science. As can be expected, these provocative statements were met with much criticism. When fellow philosophers of science (such as Agassi, 1976) rightly pointed out the absurdity of placing voodoo on a par with science, Feyerabend replied that he did not mean this in a literal sense. It was a matter of rhetoric. A nice illustration of an immunization strategy we discussed in chapter 4: setting up 'moving targets'!

SOKAL'S HOAX

So, this postmodern constructivism did not convince everyone (to say the least). Alan Sokal, a physicist and a philosopher of science, responded in a remarkable way. Not only against Feyerabend but against all postmodern thinkers who believe that there are no objective facts, only perspectives and social constructions. According to Sokal (1996a), this opens the door to a whole lot of nonsense. He believes that objective facts about the world can indeed be known and that we can and should make a distinction between meaningful, empirically supported theories and nonsensical theories about the world (i.e. that we should demarcate between science and pseudoscience).

To drive his point home, he came up with a hoax. He submitted an article to a leading academic journal 'Social Text' in the field of cultural studies and got it published (through peer review) (Sokal, 1996b). His article (entitled 'Transgressing the boundaries: Towards a transformative hermeneutics or quantum gravity') offered a strongly relativistic view of the world (as most other articles published in the postmodern journal). Sokal's article, however, was purposefully made nonsensical. It consisted mostly of grammatically correct and very esoteric sentences with many neologisms that made absolutely no sense. The hoax hit the intellectual world like a bomb! In a letter addressed to the publisher, he explained that it was an experiment to see if he would get an article past peer review that fits in with the style and philosophy of the journal, although it contained nothing but nonsense (Sokal, 1996a).

The danger of 'anything goes'

What Sokal rightly denounced is that when we open the doors of what is academically acceptable too widely, we risk drowning in nonsense.

Without a demarcation criterion, science inevitably loses its power. For two reasons. First, as we saw above, sciences can only progress if epistemic and methodological principles are shared by the scientific community - so that others can criticize and contribute to its progress. Feyerabend's 'anything goes' deprives the sciences of their greatest strength: the universal standards that allow everyone to contribute and correct each other, regardless of their personal convictions and cultural backgrounds. Secondly, we must not forget that scientists build on the work of others. If everything is admitted, including completely unreliable scientific research, then the foundations of the scientific edifice collapse.

STRIKING A BALANCE

In conclusion, I would like to offer a final piece of advice: Try to strike a balance between openness and restriction. That is the takeaway message I want to pass on to you as budding scientists in particular and as people in general. It applies both to scientific research and to our everyday thinking. We must remain open-minded and open to new and sometimes surprising ideas, but we must not open our minds to such an extent that our brains fall out! So, be open to new ideas, possibilities, perspectives, but never lose your critical gaze.

Develop the habit of gauging the reliability of a belief by considering the way it came about. Screen the arguments you are presented with for reasoning errors. And, above all, develop the habit of critically reflecting on your own thinking and beliefs. Critical thinking is an indispensable skill in the information age in which we live. It makes little sense today to cram your head full of facts that are accessible by simply consulting your smartphone. What makes sense is to develop the right filters to process that constant stream of information.

I believe critical thinking is one of the most important hiatuses in our education today and I hope that this book has filled this gap for you. Because, as I pointed out in the previous chapter, critical thinking is first and foremost a matter of responsibility. Bad thinking leads to bad outcomes. In light of the important challenges that we face today, one thing is certain: the future will be determined by the quality of our thinking. Up to you, dear student or reader, to contribute to a better world as a critical thinker!

SUMMARY

What do all (good) sciences have in common?

- They are self-correcting

Why?

- Cognitive artifacts protect against intuitive reasoning errors.
- Framework and context protect against reasoning errors of system 2 and against emotional distortion.

What is the importance of a demarcation criterion?

- Universal standards for good science ensure that scientists can criticize (and improve) each other's work.
- Scientific progress requires a reliable foundation upon which to build.

Important terms

- Dilthey's 'erklären' (explaining)
  - The aim of the natural sciences – to describe the world in terms of cause and effect and their underlying laws.
- Dilthey's 'verstehen' (understanding)
  - The aim of the social sciences – to come to a subjective understanding.
- Hacking's looping effect
  - A theory can influence its object (of inquiry) in the human sciences, since it can influence what humans think and how they behave.
- Popper's falsifiability
  - A demarcation criterion distinguishing science from pseudoscience. Scientific theories must be testable: it must be possible *in principle* to refute the theory on the basis of *observation*.

Appendix

LIST OF REASONING ERRORS

General reasoning errors

- **Confirmation bias:** The tendency to search for, interpret, favor, and recall information in a way that confirms one's preexisting beliefs or hypotheses.
- **Irrational cognitive dissonance reduction:** When information we gather from the world contradicts our beliefs, we tend to interpret that information in such a way that it no longer contradicts our beliefs.
- **Overconfidence bias:** We have too much confidence in the correctness of our own answers or beliefs.
- **Dunning-Kruger effect:** The tendency for lay people to overestimate their knowledge of something and of experts to underestimate their knowledge.
- **Bias blind spot:** We detect reasoning errors much more easily in the reasoning of others than in our own reasoning.
- **Self-overestimation:** We overestimate our own talents and prospects in life.
- **Belief bias:** Accepting the validity of an argument simply because the conclusion sounds plausible or because you agree with the conclusion.
- **Hindsight bias:** We overestimate the probability that we would have accorded to the occurrence of a certain event after the event occurred.
- **Stereotyping:** Expecting an individual of a particular group to have

certain characteristics (associated with said group) without having information about that person.

Reasoning errors of investors / consumers

- **Choice supportive bias (or post purchase rationalization):** We remember the choices we made in the past as being better than they actually were.
- **Endowment effect:** We accord more value to something simply because we own it.
- **Bandwagon effect (= ingroup bias):** We adopt beliefs too quickly when they come from people in our group and blindly follow the behavior/decisions/opinions of the group.
- **Anchoring:** A given piece of information can strongly influence our estimates (even if there is no link between that information and our estimate).
- **Framing effect:** Drawing different conclusions based on the same information because it is presented differently.
- **Loss aversion:** We feel the negative impact of a loss more intensely than the positive impact of a gain of the same size.
- **Sunk cost fallacy:** Taking into account incurred and non-recoverable costs in deciding whether to continue with a project (and thus continue to invest in it).

Statistical / mathematical reasoning errors

- **Statistical reasoning errors:** Intuitively we perform poorly at estimating probability.
- **Base rate fallacy:** We tend to ignore base rates in estimating the probability that something will occur. In general, we often turn a blind eye to general, implicit information and focus exclusively on specific, explicit information.
- **Availability bias:** We overestimate the likelihood that something will occur when it is easy to recall or imagine.

- **Gambler's fallacy:** Expecting a statistical correction when that expectation is not justified.
- **Hyperactive pattern detection:** Seeing patterns in random series.
- **Exponential reasoning errors:** We underestimate exponential growth because we are used to linear growth.

DETECT THE REASONING ERRORS

1. John is left-handed. His mother tells him frequently that left-handed people are generally more intelligent and creative than right-handed people. He is becoming more and more convinced that his mother is right because he has met numerous intelligent and creative lefthanders.
2. Joe is an avid basketball fan who likes to bet on games. He watched almost all of the NBA games for years and noticed the following: the team who scores the first point usually wins the match. Tonight, he bets on the match between the LA Lakers and the Chicago Bulls and the Lakers score the first point. So, he puts his money on the Lakers.
3. Brian studies at university and is member of a debate club where he and his fellow students analyze the ins and outs of American politics. At the first meeting they have after the election of Trump (about 5 weeks after the election), John says theatrically: 'You didn't have to be a political genius to see that Trump would be elected. You could see the frustration of white, poor Americans grow over the last decennia and Trump played right into that'. Brian is quite puzzled by John's claim because at their last meeting - 3 weeks before the election - everyone agreed unanimously that Hillary Clinton would be elected. Which bias does John succumb to?
4. Peter is a student in economics who has already saved some money. He invests his savings in shares of a biotech company. He buys the shares at 12EUR per share. After a year, the price has dropped to 8EUR per share. Peter refuses to sell his shares even though there are no indications that the share price will rise again. Which bias does he succumb to?

5. Kurt is 16 years old and just did an internship for 1 month in a garage. He mainly had to clean cars and he occasionally watched as cars were repaired. When the car of his father breaks down three months later, he is convinced that he can fix it. He sees himself as an accomplished mechanic.
6. a. Martin likes to smoke marijuana. A friend who also smokes tells him that marijuana is not harmful and even stimulates creativity. Martin decides to look it up. He consults google and types in: 'marijuana enhances creativity' and finds an endless series of websites, blogs and articles that state that marijuana indeed stimulates creativity. What reasoning error does he make here?
   b. A few years later it appears that his friend Peter, also a soft drug user, is not admitted to drama school after failing a test that requires creativity. Peter's parents send him to a psychologist. The psychologist attributes the bad performance in the creativity test to excessive marijuana use. Martin finds out about that and quickly decides that it is not so much the marijuana that made Peter fail the test, but the chronic sleep deprivation by playing video games all night (albeit with a joint). What reasoning error does he make here?
7. You want to buy a new mobile phone. After looking around in the store for a long time, assisted by a diligent salesperson, you decide to buy model X. You are very happy with the purchase and proudly show the new phone to your friends. A friend asks you if that model has the new panoramic camera function. You quickly say 'yes', but you don't really know. You hope so. In the evening you take the manual and discover that there is no panoramic function. 'Damn it," you yell out in distress, but soon you figure that you do not really need it and that this panoramic function is a sales trick anyway. Everyone knows that you cannot take beautiful panoramic photos with mobile phones.
8. A marketing agency advises the sales department of Tesla to exaggerate how many people have already ordered a car. Which bias do they hope to exploit?

9. John is an American who does not believe climate change is happening. Recently he heard a Republican governor say: 'Last month of August I had to put on a sweater almost every single night. Imagine ... In the middle of the summer! Therefore, climate change cannot be happening." John thought that this was a strong and valid argument and now uses it himself in discussions on climate change.
10. A new biotech company announces that it will bring a fantastic product to the market. The product gets a lot of attention in the media. It attracts a huge amount of investment, even though only 20% of starting biotech companies ever become profitable.
11. The Supermarket 'Albert Hein' has bought a new Chilean wine and hopes to make a nice profit selling it. They buy the wine for 2,5EUR per bottle and want to sell it for 6EUR per bottle. But the wine sells poorly. A marketing agency advises them to put the Chilean wine on the racks between their cheapest (and not very nice-looking bottles) and their much more expensive (and for most people unaffordable) bottles. Which bias are they trying to exploit here?
12. a. Linda and Mary begin a start-up. At the start they both invest 20 000EUR in the project. After the first year, that money is completely spent (they have rented and furnished an office space, made marketing costs, etc.). Unfortunately, the start-up still hasn't generated any returns. If they do not pump in extra capital, it will be game over. They decide to ask their parents for extra funds because they do not want to give up now, given that they have already invested so much effort and money. What reasoning error do they make?
    b. Moreover, they add, we have read about a lot of start-ups that also faced difficult times in the beginning and are now very successful. What reasoning error do they make here?
13. 5% of lifelong smokers get lung cancer. Recently an article was published in which the CEO of Philip Morris talks about the unjustified demonization of the tobacco industry. He stressed that tobacco is a leisure product and added that the vast majority of

smokers - 95% - do not get lung cancer. Which bias does the CEO exploit here?

14. A friend of yours bought a ticket to a concert but cannot go. He asks you if you would buy it from him. You decide that you want to spend no more than 30 EUR. Your friend accepts and you buy the ticket for 30 EUR. A day before the concert someone offers you 40 EUR for the ticket. You refuse to sell it.
15. Sabine already has three children and is pregnant with her fourth. Her first three children are girls. She asks her two sisters to guess the gender. Sandra, Sabine's eldest sister, thinks it will be a boy. Because four girls in a row seems very improbable.
16. There is a new machine on the market that detects counterfeit money. A large marketing campaign is set up to sell the machine to stores. The campaign boasts that the machine detects 99.999% of fake money. That seems very accurate. Many shops therefore buy the machine. What, however, have they forgotten to check and what kind of reasoning error do they make?
17. Jeremy believes that women are bad drivers. The longer he has been driving, the more convinced he becomes of this because he has seen so many bad female drivers over the years.
18. a. Kurt believes that eating gluten is harmful and he is talking about it with his girlfriend Ann. He tries to convince her by pointing out that two of his friends have recently stopped eating gluten and reportedly feel much better. Ann took a course on critical thinking and points out that he is making a reasoning error. What reasoning error does Kurt make?
    b. Ann explains the reasoning error he made, but Kurt refuses to see that something is wrong with his reasoning. 'His thinking', he tells Ann, 'is always rational'. 'In contrast to many others,' he adds. What reasoning error does Kurt make here?
19. A long time ago there was an Indian Maharaja who loved to play chess. He was always looking for new opponents. To encourage people to play, he promised them a prize if they could win. Usually, it was a copper cup or a necklace for their wife. One day, a beggar came to the Maharaja to play chess. The Maharaja promised him a

cup that he would receive if he won. The beggar, however, turned down the offer and said: 'Honorable Maharaja, the only thing I want is a little rice. If I win, do you agree to put 1 grain of rice on the first square of the chessboard and then double the number in the next square (i.e. 2,4,8, etc.) until the whole chessboard is filled? 'The Maharaja agreed and thought he would be off the hook by giving a little bit of rice if he lost. What reasoning error does the Maharaja make? (Source: https://www.mathscareers.org.uk/the-rice-and-chessboard-legend/)

ANSWERS

1. **Self-overestimation and the confirmation bias**
   Self-overestimation: We overestimate our talents and prospects in life.
   John overestimates his creativity as a left-handed person.
   Confirmation bias: The tendency to search for, interpret, favor, and recall information in a way that confirms one's preexisting beliefs or hypotheses.
   John only sees and remembers confirmation for his belief: smart and creative left-handed people. This strengthens his conviction.
2. **Hyperactive pattern detection, the confirmation bias and the overconfidence bias**
   Hyperactive pattern detection: Seeing patterns in random series. There is no relationship between scoring the first point and winning the game, Joe is mistaken in thinking there is one. He sees a pattern that is not there.
   Confirmation bias: Over the years Joe has seen and remembered mostly confirming instances of his belief (games won by teams who scored the first point) and has not noticed or quickly forgotten the disconfirming instances (games lost by teams who won the first point).
   Overconfidence bias: We have too much confidence in the correctness of our own answers (and predictions).
   Joe overestimates the likelihood that he will predict the outcome of the game correctly and therefore goes 'all in'.
3. **Hindsight bias:** We overestimate the probability that we would have accorded to the occurrence of a certain event after the event actually occurred.
   Brian reasons afterwards that (before the elections) he predicted the victory of Trump, but that was actually not the case.
4. **Loss aversion:** We feel the negative impact of a loss more intensely than the positive impact of a gain of the same size.
   Peter's loss feels uncomfortable to him, so he tends not to sell his

shares below the purchase price, although he would better do so, since there is no indication that the share price will go up again.

5. **Dunning-Kruger effect:** The tendency for lay people to overestimate their knowledge of something and for experts to underestimate their knowledge.
   Kurt is clearly no expert and overestimates his own knowledge.
6. a. **Confirmation bias:** Martin only looks for confirmation of his beliefs.
   b. **Irrational cognitive dissonance reduction:** When information we gather from the world contradicts our beliefs, we tend to interpret that information in such a way that it no longer contradicts our beliefs.
      Martin receives information that refutes his conviction but tries to save his conviction from falsification by giving a different interpretation to the facts.
7. **Choice supportive bias or post purchase rationalization:** We remember the choices we made in the past as being better than they actually were.
   Because (without knowing it) you bought a phone without a panoramic function, you argue that you don't want that function anyway (even though you would have liked that function). You rationalize that you made a good purchase, when in fact you did not.
8. **Bandwagon effect:** We adopt beliefs too quickly when they come from people in our group and blindly follow the behavior/decisions/opinions of the group.
   The advertisement wants to make it appear that many people in your own group already bought this car hoping that you will jump on the bandwagon and follow suit.
9. **Belief bias:** Accepting the validity of an argument simply because the conclusion seems plausible or because you agree with the conclusion.
   John adopts a bad argument uncritically (based on a subjective perception of temperature over a short period of time, we cannot determine global climate trends), because he agrees with the conclusion.

10. **Base rate fallacy:** We tend to ignore base rates in estimating the probability that something will occur. In general, we often turn a blind eye to general, implicit information and focus exclusively on specific, explicit information.
    The investors do not or barely take into account the fact that only 20% of biotech start-ups ever become profitable (that 20% is the base rate).
11. **Anchoring:** A given piece of information can strongly influence our estimates (even if there is no link between that information and our estimate). (Anchoring is a form of framing).
    When the wine is presented by itself, people see it as too expensive for the quality of the wine (since the wine does not sell well). The hope is that if you put the wine next to a very expensive bottle, it will seem affordable, and by putting a very cheap-looking bottle next to it, it will seem qualitative.
12. a. **Sunk cost fallacy:** Taking into account incurred and non-recoverable costs in deciding whether or not to continue with a project (and thus continue to invest in it).
       Linda and Mary take their decision to continue based on the fact that they have already invested a lot of money. This is irrational since they should only be concerned with the expected profits of the new investment.
    b. **Availability bias:** We overestimate the likelihood that something will occur when it is easy to recall or imagine.
       Because Linda and Mary have read a lot about successful start-ups (the only ones that are covered in the press), they overestimate the chance that start-ups (even with a difficult start) succeed and assume that this will also be the case for their start-up.
13. **Framing effect:** Drawing different conclusions based on the same information because it is presented differently.
    The CEO expresses the statistics in such a way (95% do not get lung cancer) to paint the picture that the health risks of smoking are not that bad. Saying that 1 in 20 smokers will get lung cancer (which of course amounts to the same thing) would sound much

more alarming and therefore might convince more people to stop smoking.

14. **Endowment effect:** We accord more value to something simply because we own it.
    Before you had the ticket, you thought it was worth a maximum of 30 EUR, now that you have it in your possession, it is suddenly worth more than 40 EUR to you.
15. **Gambler's fallacy:** Expecting a statistical correction when that expectation is not justified.
    The chance of having a girl or a boy is the same every time. The previous births of girls do not make it more likely that it will be a boy this time. There's no justification for expecting a statistical correction.
16. **Statistical reasoning error:** Intuitively we perform poorly at estimating probability (or at statistical reasoning).
    They have forgotten to check how much real money the machine actually detects as real. You could make a machine that says 'false' with every note, and therefore will be 100% accurate in detecting counterfeit money.
17. **Stereotyping and the confirmation bias**
    Stereotyping: Expecting an individual of a particular group to have certain characteristics (associated with that group) without having information about that person.
    Jeremy erroneously expects every woman to meet his stereotype.
    Confirmation bias: (see above)
    He is more and more convinced that women are bad drivers because he is much more receptive to confirming instances (women driving badly) and remembers such instances much better than disconfirming instances (women driving well / men driving badly).
18. a. **Confirmation bias:** (see above)
       Ann makes it clear to Kurt that he should not only be looking for confirmation of his belief (the two friends who feel better on a gluten-free diet), but also for possible counterevidence.
    b. **Bias blind spot:** We detect reasoning errors much more easily in others than in ourselves.

Kurt is blind to his own reasoning error and thinks that others make more reasoning errors than he does himself.

19. **Exponential reasoning error:** We underestimate exponential growth because we are used to linear growth.
The Maharajah underestimates how much rice he should put on the chessboard because he underestimates exponential growth. On the 64th and final chess square alone, the Maharajah should lay 18 000 000 000 000 grains of rice. That is more than 210 billion tons of rice. With that amount of rice you can cover the whole of India one meter deep and it is much more rice than has been produced throughout the history of the world!

Bibliography

Agassi, J. (1976). Review of against method. *Philosophia*, *6*, 165–177. https://doi.org/10.1007/BF02383263 (see p. 100).

Allais, M. (1953). Le comportement de l'homme rationnel devant le risque: critique des postulats et axiomes de l'école Américaine. *Econometrica : journal of the Econometric Society*, *21*(4), 503–546. https://doi.org/10.2307/1907921 (see p. 21).

Ariely, D. (2008). *Predictably irrational: The hidden forces that shape our decisions*. Harper Perennial. (See pp. 13, 24, 57, 60).

Barrett, J. (2000). Exploring the natural foundations of religion. *Trends in cognitive sciences*, *4*(1), 29–34. https://doi.org/10.1016/S1364-6613(99)01419-9 (see pp. 33, 55).

Blainey, G. (1988). *The causes of war*. Free Press. (See p. 78).

Bostrom, N. (2013). Existential risk prevention as global priority. *Global Policy*, *4*, 15–31. https://doi.org/10.1111/1758-5899.12002 (see p. 96).

Boudry, M. (2016). *Illusies voor gevorderden*. Uitgeverij Polis. (See pp. 11, 52, 78).

Braeckman, J. (2017). *Valkuilen van ons denken. Een hoorcollege over de kracht van kritisch denken*. Home Academy. (See p. 13).

Braeckman, J., & Boudry, M. (2011). De ongelovige Thomas heeft een punt. *Houtekiet*. (See pp. 47, 48).

Casler, K., & Kelemen, D. (2008). Developmental continuity in teleo-functional explanation: Reasoning about nature among Romanian Romani adults. *Journal of Cognition and Development*, *9*(3), 340–362. https://doi.org/10.1080/15248370802248556

Clark, A., & Chalmers, D. (1998). The extended mind. *Analysis*, *58*(1), 7–19. https://doi.org/10.1093/analys/58.1.7 (see p. 72).

Cosmides, L., & Tooby, J. (1992). *Cognitive adaptations for social exchange* (J. Barkow, L. Cosmides, & J. Tooby, Eds.). Oxford University Press. (See p. 65).

Darwin, C. (1958). *The autobiography of Charles Darwin 1809-1882* (N. Barlow, Ed.). W.W. Norton & Company. (Original work published 1887). (See p. 51)

Dawkins, R. (1976). *The selfish gene*. Oxford University Press.

Dawkins, R. (1986). *The blind watchmaker*. W. W. Norton & Company.

De Cruz, H., & De Smedt, J. (2007). The role of intuitive ontologies in scientific understanding – the case of human evolution. *Biology \& Philosophy*, *22*, 351–368. https://doi.org/10.1007/s10539-006-9036-8 (see p. 63).

De Regt, H., & Dooremalen, H. (2008). Wat een onzin! Wetenschap en het paranormale. (See p. 13).

De Regt, H., & Dooremalen, H. (2015). Het snapgevoel: Hoe de illusie van begrip ons denken gijzelt. (See p. 13).

Dilthey, W. (1989). *Introduction to the human sciences*. Princeton University Press. (Original work published 1883). (See p. 90)

Dooremalen, H. R., H., S., & M. (2021). Exploring humans: Philosophy of science for the social sciences – A historical introduction. (See p. i).

Erickson, T., & Mattson, M. (1981). From words to meaning: A semantic illusion. *Journal of Verbal Learning & Verbal Behavior*, *20*(5), 540–551. https://doi.org/10.1016/S0022-5371(81)90165-1 (see p. 30).

Festinger, L. (1957). *A theory of cognitive dissonance*. Stanford University Press. (See p. 69).

Feyerabend, P. (1970). *Against method: Outline of an anarchistic theory of knowledge*. University of Minnesota Press. (See pp. 99, 100).

Fischhoff, B. (1975). Hindsight is not equal to foresight: The effect of outcome knowledge on judgment under uncertainty. *Journal of Experimental Psychology: Human Perception and Performance*, *1*, 288–299. https://doi.org/10.1037/0096-1523.1.3.288 (see p. 23).

Fischhoff, B., Slovic, P., & Lichtenstein, S. (1977). Knowing with certainty: The appropriateness of extreme confidence. *Journal of Experimental Psychology: Human Perception and Performance*, *3*, 552–564. https://doi.org/10.1037/0096-1523.3.4.552 (see p. 71).

Gigerenzer, G. (2000). *Adaptive thinking: Rationality in the real world*. Oxford University Press. (See p. 64).

Gilovich, T., Tversky, A., & Vallone, R. (1985). The hot hand in basketball: On the misperception of random sequences. *Cognitive Psychology*, *17*(3), 295–314. https://doi.org/10.1016/0010-0285(85)90010-6 (see p. 46).

Gladwell, M. (2005). *Blink: The power of thinking without thinking*. Little, Brown and Co. (See p. 67).

Goldacre, B. (2008). *Bad science*. Fourth Estate. (See p. 60).

Hacking, I. (1995). The looping effects of human kinds. In D. Sperber, D. Premack, & A. Premack (Eds.), *Causal cognition: A multidisciplinary debate* (pp. 351–394). Clarendon Press/Oxford University Press(see p. 91).

Haselton, M., & Buss, D. (2000). Error management theory: A new perspective on biases in cross-sex mind reading. *Journal of Personality and Social Psychology*, *78*(1), 81–91. https://doi.org/10.1037/0022-3514.78.1.81 (see p. 32).

Hook, S. (1959). *Psychoanalysis, scientific method, and philosophy a symposium*. New York University Press. (See p. 51).

Johnson, D. (2004). *Overconfidence and war: The havoc and glory of positive illusions*. Harvard University Press. (See p. 79).

Kahneman, D. (2011). *Thinking, fast and slow*. Farrar, Straus and Giroux. (See pp. 20, 30, 34, 41, 61).

Kahneman, D., & Frederick, S. (2002). Representativeness revisited: Attribute substitution in intuitive judgment. (See p. 31).

Kahneman, D., & Tversky, A. (1979). Prospect theory: an analysis of decision under risk. *Econometrica: Journal of the Econometric Society*, *47*(2), 263–291. https://doi.org/10.2307/1914185 (see p. 21).

Kahneman, D., & Tversky, A. (1985). *Evidential impact of base rates* (D. Kahneman, P. Slovic, & A. Tversky, Eds.). Cambridge University Press. (See p. 16).

Kelemen, D. (2003). British and American children's preferences for teleo-functional explanations of the natural world. *Cognition*, *88*(2), 201–21. https://doi.org/10.1016/S0010-0277(03)00024-6

Kuhn, T. (1970). *The structure of scientific revolutions*. University of Chicago Press. (See p. 2).

Lord, C., Ross, L., & Lepper, M. (1979). Biased assimilation and attitude polarization: The effects of prior theories on subsequently considered evidence. *Journal of Personality and Social Psychology*, *37*(11), 2098–2109. https://doi.org/10.1037/0022-3514.37.11.2098 (see p. 37).

Malkiel, B. (2003). *A random walk down Wall Street : The time-tested strategy for successful investing*. W.W. Norton. (See p. 80).

Norenzayan, A. (2013). *Big gods: How religion transformed cooperation and conflict*. Princeton University Press. (See p. 81).

Okasha, S. (2002). *Philosophy of science: A very short introduction*. Oxford University Press. (See pp. 1, 8).

Pinker, S. (1997). *How the mind works*. Norton. (See p. 34).

Popper, K. (1963). *Conjectures and refutations: The growth of scientific knowledge*. Routledge & K. Paul. (See pp. 2, 49, 50, 60, 98).

Proust, M. (1949). *A la recherche du temps perdu*. Gallimard. (See p. 77).

Raiffa, H. (2002). *Negotation analysis: The science and art of collaborative decision making*. Harvard University Press. (See p. 16).

Sagan, C. (1980). *Cosmos*. Random House. (See p. 93).

Singer, P. (1993). *How are we to live? Ethics in an age of self-interest*. Oxford University Press. (See p. 83).

Sokal, A. (1996a). A physicist experiments with cultural studies. *Lingua Franca*, *6*(4), 62–64. (See p. 101).

Sokal, A. (1996b). Transgressing the boundaries: Toward a transformative hermeneutics of quantum gravity. *Social Text*, *46*(47), 217–252. https://doi.org/10.2307/466856 (see p. 101).

Sperber, D., & Mercier, H. (2017). *The enigma of reason. A new theory of human understanding*. Allen Lane & Harvard University Press. (See pp. 36, 41, 61).

Surowiecki, J. (2004). *The wisdom of crowds: Why the many are smarter than the few and how collective wisdom shapes business, economies, societies, and nations*. Doubleday & Co. (See p. 71).

Tomasello, M. (2009). *Why We Cooperate*. MIT Press. (See p. 73).

Tooby, J., & Cosmides, L. (1992). The psychological foundations of culture. In J. Barkow, L. Cosmides, & J. Tooby (Eds.), *The adapted mind: Evolutionary psychology and the generation of culture* (pp. 19–136). Oxford Univ. Press(see p. 33).

Tversky, A., & Kahneman, D. (1973). Availability: A heuristic for judging frequency and probability. *Cognitive Psychology*, *5*(2), 207–232. https://doi.org/10.1016/0010-0285(73)90033-9 (see p. 19).

Tversky, A., & Kahneman, D. (1974). Judgment under uncertainty: Heuristics and biases. *Science (New York, N.Y.)*, *185*(4157), 1124–1131. https://doi.org/10.1126/science.185.4157.1124 (see p. 64).

Tversky, A., & Kahneman, D. (1983). Extensional versus intuitive reasoning: The conjunction fallacy in probability judgment. *Psychol. Rev*, *90*, 4. https://doi.org/10.1037/0033-295X.90.4.293 (see p. 15).

Vlerick, M. (2019). *De tweede vervreemding: Het tijdperk van de wereldwijde samenwerking*. Lannoo. (See p. 5).

Wason, P. (1960). On the failure to eliminate hypotheses in a conceptual task. *Quarterly Journal of Experimental Psychology*, *12*, 129–140. https://doi.org/10.1080/17470216008416717 (see p. 23).

Made in United States
Orlando, FL
05 November 2026

90230158R00077